AIR
QUALITY
PERMITTING

R. Leon Leonard

LEWIS PUBLISHERS

Boca Raton   New York   London   Tokyo

Acquiring Editor:	Ken McCombs
Project Editor:	Les Kaplan
Marketing Manager:	Greg Daurelle
Cover design:	Dawn Boyd
PrePress:	Kevin Luong
Manufacturing:	Sheri Schwartz

Library of Congress Cataloging-in-Publication Data

Leonard, R. Leon
    Air quality permitting / R. Leon Leonard.
       p.   cm.
    Includes bibliographical references and index.
    ISBN 0-87371-790-2
       1. Air--Pollution--Law and legislation--United States.
    2. Environmental permits--United States. I. Title
    KF3812.L46 1996
    344.73'046342--dc20
    [347.30446342]96-20443

              96-20443
              CIP

PREFACE

The purpose of this book is to show facility air quality managers, regulatory agency staff, and consultants how to prepare permit applications for Title V Federal Operating Permits or for New Source Review (NSR) or Prevention of Significant Deterioration (PSD) permits for new or modified sources of air pollutant emissions, and how to comply with permits obtained.

To successfully obtain air quality permits for a facility, permit preparers must develop permitting and compliance strategies that take into account all of the regulatory requirements and how they can be satisfied. This book provides that information. It suggests strategies that will allow the user to move through the regulatory process as quickly and cost effectively as possible, and to obtain a permit that ensures compliance with applicable regulations, without unnecessary burdens. The book will also be useful to consultants who are asked to assist facilities in permitting, because it brings together all the elements needed for permit application preparation in one place. Finally, the book will aid regulatory staff in understanding the air permitting process from the perspective of the applicant, thereby enabling them to carry out their responsibilities for improving air quality and protecting public health.

Chapter 1 reviews the regulatory requirements that must be met by an emission source seeking a permit, including thresholds for permitting, prohibitions, and the requirements for federal Title V operating permits imposed by the Clean Air Act Amendments of 1990.

Chapter 2 shows the reader how to estimate emissions from air pollutant sources using AP-42 and other emission factors, material balances, engineering calculations, and source testing. The difficult area of fugitive emission estimates is also addressed.

Chapter 3 describes what is required to satisfy Best Available Control Technology (BACT) requirements, the "top down" analysis required by EPA, and other air quality control technology standards. Maximum Available Control Technology (MACT) requirements for sources of hazardous air pollutants are also discussed.

Emission reduction credits, reductions in emissions from existing sources, "netting out," and emission offsets are discussed in Chapter 4. This discussion includes the methods required for determining what credits can be counted, existing source reduction methods, discounts, and offset ratios.

Modeling to determine the air quality impacts of emission sources is discussed in Chapter 5. The information needed to oversee air quality modeling and ensure its quality is clearly provided.

Chapter 6 discusses permitting requirements that are imposed for air toxics, including hazardous air pollutants. Assumptions and methods used to calculate health risk are provided, and new developments in health-risk assessments, such as stochastic modeling, are examined. Technology-based MACT standards and resultant National Emission Standards for Hazardous Air Pollutants (NESHAP) are also reviewed.

Permitting strategy development is discussed in Chapter 7. A facility must develop a strategy for preparation of the air permit application that is consistent with both existing and anticipated regulatory requirements and that includes consideration of the personnel, time, and cost of permit preparation.

Once a facility has received a permit, attention must be given to compliance with permit conditions and enforcement, as discussed in Chapter 8. This issue has grown in importance as larger fines and even criminal penalties are being imposed by regulatory agencies.

This book has been developed while the author has worked permitting new air emission sources and evaluating permit compliance in nineteen states. It evolved while the author taught a course on Air Quality Permitting for the University of California at Davis Extension Certificate Program in Air Quality Management.

R. Leon Leonard, Ph.D.
Bellevue, Washington

ACKNOWLEDGMENTS

I wish to acknowledge deeply the help and constant support of my wife, Janet, and her continuing prayers while I completed this work.

In addition, I want to acknowledge the following people:

Nancy Hill, who typed the first draft. I mourn her untimely death.

Mona Ellerbrock, who gave me early encouragement.

Ken Selover, a continuing friend and professional colleague, who contributed to Chapter 7.

Gary Lucks, who provided comments on Chapter 7.

Mark Ludwiczak, who provided comments on source testing in Chapter 2.

Wyatt Dietrich, who gave persistent gentle encouragement by repeatedly asking, "How's the book going?"

Lucy Trumbull and Jim Hunt, who contributed graphic creativity in preparing most of the figures in the text.

Pat Nelson, who provided examples and perspective on BACT that I used in Chapter 3.

Gregg Spadorcio and Toni Hardesty, who provided funding for graphics and table preparation.

Sharon Juen, who carefully formatted all the tables.

Russ Henning, who provided answers to countless questions about material I'd forgotten because I've been out of school so long, or that didn't exist yet when I was in school.

The staff at the University of California at Davis Hazardous Material Management and Air Quality Management Certificate Programs, for their patience with my late handouts and other foibles.

The staff at Radian International LLC, who have provided consistent professional support, frequent challenges, and lots of new ideas to explain how to do air quality permitting.

All the other people who are now slighted that I didn't mention them by name, in spite of the numerous times they have rescued me from my own ineptitudes. I would have included you, but doing so would have reminded me too much of the ineptitudes. Thank you all very much.

CONTENTS

CHAPTER 2

CHAPTER 3

CHAPTER 6

1 REGULATORY REQUIREMENTS FOR PERMITTING

I. INTRODUCTION

The purpose of air quality permitting is to regulate the emissions of air pollutants from sources within the jurisdiction of the permitting agency. Air pollution control agencies are responsible for protecting the quality of air within their jurisdictions, and for improving the air quality if it does not currently meet the ambient standards.

A. Permitting Authority

The Environmental Protection Agency (EPA) is responsible for establishing and maintaining federal programs to control air quality, including mobile source standards. Each of the states is responsible for air quality within its borders, although this responsibility may be shared with Native Americans on tribal lands. In many states, jurisdiction has been delegated to regional or local agencies responsible for air quality in a particular air basin.[1] Air basins, although their boundaries are defined physically by terrain relief or water bodies, usually are designated by county boundaries for regulatory convenience.

B. Delegation of Authority

EPA has the authority for assuring compliance with the Clean Air Act. Some air quality programs, such as New Source Performance Standards (NSPS), and federal rule making are carried out directly by EPA. Where a federal Prevention of Significant Deterioration (PSD) or New Source Review (NSR) permit must be issued by EPA, it is the responsibility of one of the ten geographical EPA regions. However, in most states, programs have been delegated by EPA, and in some states, the delegation is to a regional or local agency within the state. This delegation is not uniform. EPA has evaluated the regulatory programs proposed by state and local agencies to determine whether they meet the requirements of federal law for that program. If the program is *at least as stringent* as the federal law requires, EPA can delegate the program to the state or regional agency. Care should be taken in seeking an air quality permit to be sure that the agency granting the permit has received delegation from EPA for the permit program involved. If the agency has not been delegated authority, it probably will be necessary to obtain an additional permit from EPA.

For large sources of air pollutants, permits are required by the permitting agency responsible for the geographic area in which the source is located. In some jurisdictions, small sources do not require permits.

Historically, permits may also have not been required for sources that existed before the agency required permits. These sources are termed *grandfathered sources*. The degree to which grandfathered sources escape permitting has varied among states, but as more stringent laws and regulations have been imposed to address nonattainment of the ambient air quality standards, fewer and fewer sources have retained a grandfathered, unpermitted, uncontrolled status. Under the provisions of Title V of the Clean Air Act Amendments of 1990 (CAAA), all major sources of air pollutants are required to have federal Title V air permits. Title V permits are discussed below.

C. Permit Processing Steps

A diagram of the steps in the permitting process is shown in Figure 1-1. The first step is to define the project. Project definition may not be complete at the time permitting is begun; when this is the case, the applicant will be at some risk. This is discussed under permitting strategy in Chapter 7.

The second step in the permitting process is to estimate the emissions from the proposed project. Both criteria and toxic pollutant emissions must be estimated. Techniques for emission estimation are described in Chapter 2.

Once emission rate estimates are available, the permitting agency regulations should be reviewed to determine whether the project emissions exceed thresholds for permitting. The regulations also should be reviewed for control requirements such as best available control technology (BACT), discussed in Chapter 3; offsets, discussed in Chapter 4; air quality impact analysis (Chapter 5); and toxic air pollutant requirements (Chapter 6). These thresholds are specified for both criteria and toxic air pollutants. Thresholds are discussed below.

If the BACT threshold is exceeded, emission controls for the source must be BACT. The determination of what technology or emission control requirement constitutes BACT is described in Chapter 3. After the BACT analysis is complete, it will be necessary to recalculate the emission rates incorporating the BACT level of control. More stringent control requirements are imposed if the source is in a nonattainment area.

If the offset threshold is exceeded, emission reductions from existing sources must be imposed to offset the increases from the proposed source. The require-

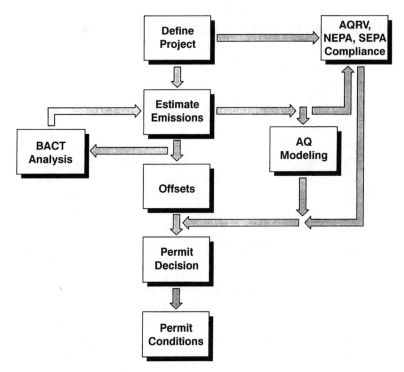

Figure 1-1 Air quality permitting steps.

ments for these emission reductions, and strategies for locating and achieving such reductions, are discussed in Chapter 4.

For Prevention of Significant Deterioration (PSD) permits and in some other circumstances, it may be necessary to demonstrate that changes in ambient concentrations of pollutants that are caused by emissions from the proposed source will not exceed PSD increments or the ambient air quality standards. This requires the use of computer-based air dispersion models. The requirements for this modeling are described in Chapter 5.

Additional impacts may need to be addressed under the provisions of air quality related values (AQRV) in PSD permitting, or the National Environmental Policy Act (NEPA) or state equivalents. AQRV requirements are discussed as part of the PSD program below, and NEPA requirements are discussed in Chapter 7.

Compliance with air quality permits and the enforcement of permits are discussed in Chapter 8.

D. Ambient Air Quality Standards

Air quality in the United States is determined for six common pollutants by measurements of the concentration of these pollutants in ambient air (e.g., air located in areas of public access).

THERE ARE SIX FEDERAL CRITERIA POLLUTANTS

- Ozone (O_3)
- Carbon monoxide (CO)
- Nitrogen dioxide (NO_2)
- Sulfur dioxide (SO_2)
- Inhalable particulate matter (PM_{10})
- Lead (Pb)

For each of these pollutants, research conducted on their health effects has been compiled by the EPA; concentrations of each pollutant above which adverse health effects will occur have been established. These concentrations are the *criteria* for health effects for each pollutant. Because ambient air quality criteria have been established, these six pollutants are termed *criteria pollutants.** As part of each ambient air quality criterion, there is a time period for which that concentration is acceptable. The longer the exposure to a pollutant, the lower the concentration that will produce adverse health effects. Consequently, the criteria for long averaging times are lower than those for shorter averaging times. The list of criteria and their averaging times for each of the six criteria pollutants are termed the *ambient air quality standards*. Federal ambient air quality standards and California state standards are shown in Table 1-1. Other states have also adopted more stringent standards.

AMBIENT STANDARDS HAVE SEVERAL CHARACTERISTICS

- Primary and secondary standards
- Ambient concentration includes
  - Units of parts per million or $\mu g/m^3$
  - Averaging time for each standard
  - Specified significant digits
  - Exceedence criterion

Table 1-1 identifies two sets of federal standards: primary standards and secondary standards. The primary standards are criteria based upon levels and averaging times that cannot be exceeded without affecting human health. The secondary standards are based upon levels and averaging times that cannot be exceeded without affecting human welfare and the environment. Although the

---

* The most recent addition to the criteria pollutant list is *particulate matter less than 10 μm in aerodynamic diameter (PM_{10})*. It became a criteria pollutant on July 1, 1987 by being substituted for a prior criteria pollutant, *total suspended particulates (TSP)*. The earliest research on particulates had not clearly distinguished the effect of the size of particles on health effects. However, later work identified that the only particles that were able to penetrate to the smaller passages in the lungs where they could be absorbed and impact health were very small. Measurements differed on just how small a particle needed to be before it would affect health, but there were enough data to suggest that effects were observed for particles below 10 μm in size for EPA to establish a 10-μm criterion. Current research suggests that health effects are only caused by even smaller particles, so there is discussion about establishing a 2.5-μm criterion.

Table 1-1 Ambient Air Quality Standards

Pollutant	Symbol	Averaging time	Standard California	Standard National	Exceedence Criteria California	Exceedence Criteria National
Ozone	O_3	1 hr	0.09 ppm (180 µg/m³)	0.12 ppm (235 µg/m³)	If exceeded	If exceeded on more than 3 days in 3 years
Carbon monoxide	CO	8 hr	9.0 ppm (10 mg/m³)	9 ppm (10 mg/m³)	If exceeded	If exceeded more than 1 day per year
(Lake Tahoe only)		8 hr	6 ppm (7 mg/m³)	—	If exceeded	
		1 hr	20 ppm (23 mg/m³)	35 ppm (40 mg/m³)	If exceeded	If exceeded more than 1 day per year
Inhalable particulate matter	PM_{10}	Annual geometric mean	30 µg/m³		If exceeded	
		Annual arithmetic mean		50 µg/m³		If exceeded
		24 hr	50 µg/m³	150 µg/m³	If exceeded	If exceeded more than 1 day per year
Nitrogen dioxide	NO_2	Annual average		0.053 ppm (100 µg/m³)		If exceeded
		1 hr	0.25 ppm (470 µg/m³)		If exceeded	
Sulfur dioxide	SO_2	Annual average		80 µg/m³ (0.03 ppm)		If exceeded
		24 hr	0.04 ppm (105 µg/m³)	365 µg/m³ (0.14 ppm)	If exceeded	If exceeded more than 1 day per year
(Secondary federal standard)		3 hr		1300 µg/m³ (0.5 ppm)		If exceeded more than 1 day per year
		1 hr	0.25 ppm (655 µg/m³)		If exceeded	
Lead particles	Pb	Calendar quarter		1.5 µg/m³		If exceeded
		30 days	1.5 µg/m³		If equaled or exceeded	
Sulfate particles	SO_4	24 hr	25 µg/m³	—	If equaled or exceeded	

Table 1-1 Ambient Air Quality Standards (Continued)

Pollutant	Symbol	Averaging time	Standard California	Standard National	Exceedence Criteria California	Exceedence Criteria National
Hydrogen sulfide	H$_2$S	1 hr	0.03 ppm (42 µg/m³)	—	If equaled or exceeded	
Vinyl chloride	C$_2$H$_3$Cl	24 hr	0.010 ppm (26 µg/m³)		If equaled or exceeded	

Source: California Air Resources Board

Notes: All standards are based on measurements at 25°C and 1 atmosphere pressure. Decimal places shown for standards reflect the rounding precision used for evaluating compliance. National standards shown are the primary (health effects) standards, except as noted. Concentration expressed first in units in which it was promulgated.

terminology *secondary standard* suggests a lesser importance, both the primary and secondary standards are levels that cannot be exceeded.

Another distinction among criteria pollutants is in how they are created (Figure 1-2). Primary pollutants are those emitted directly by sources; all of the criteria pollutants except ozone and PM_{10} are primary. Ozone is a secondary pollutant because it is created by a reaction in the atmosphere between its *precursors*, nitrogen oxides and volatile organic compounds (VOC).* PM_{10} is both a primary and secondary pollutant. It can be emitted directly from sources. It can also be created from oxidation in the atmosphere of sulfur dioxide (SO_2) to sulfates and nitrogen oxides (NO_x) to nitrates. Thus SO_2 and NO_x are considered precursors of PM_{10}. A third source of PM_{10} is high molecular weight hydrocarbons, which can condense to become liquid particles after having been emitted as gases at higher temperature.

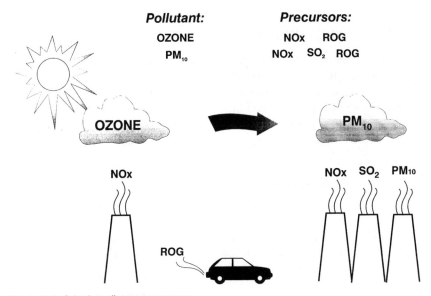

Figure 1-2 Criteria pollutant precursors.

E. Attainment and Nonattainment Areas

If the concentration of a criteria pollutant in a geographical area is below all of the ambient air quality standards for that pollutant, the area is an *attainment area* for that pollutant. If any of the ambient air quality standards is exceeded, that area is considered *nonattainment*.** More than 20 urban areas are currently

---

* Not all hydrocarbons react in the atmosphere to form ozone. Methane, ethane, acetone, and chlorinated hydrocarbons are excluded. Very heavy organic compounds are not considered volatile, and therefore are excluded. The distinctions among hydrocarbons that do and do not react in the atmosphere to form ozone has resulted in several different terms found in the regulations of different agencies. The term used by EPA is *volatile organic compounds* (VOC), but you will also see *reactive organic gases* (ROG), *volatile organic matter* (VOM), and other variants.

** Attainment is determined for each pollutant separately. The standards referred to here are those for the different averaging times for each pollutant (e.g., for SO_2, there are 3-hr, 24-hr, and annual averaging time standards, and SO_2 is a nonattainment pollutant if any of these standards is exceeded in the area.

considered nonattainment for ozone. More than 80 are nonattainment for carbon monoxide. More than 15 areas are nonattainment for sulfur dioxide, and hundreds of areas are nonattainment for PM_{10}.[1]

It is a statutory goal of EPA that areas that are nonattainment for any of the criteria pollutants will develop State Implementation Plans (SIPs) to identify how those areas will manage the control of air pollutants to bring them into attainment. The SIPs must identify a strategy, consisting of rules limiting emissions of nonattainment pollutants and their precursors, and a time frame for reaching attainment. Because of these rules, the requirements that must be met by a new or modified source of air pollutants in a nonattainment area are more stringent than those required for new or modified sources in attainment areas. Consequently, most of the remaining discussion in this chapter will distinguish between the permitting requirements in attainment areas and those in nonattainment areas.

II. SOURCES NEEDING PERMITS

Stationary sources of air pollutant emissions usually require permits. Several thresholds to permit requirements are discussed in the next section. If permitting thresholds are exceeded, either in a new or modified source, a permit is usually required before the source or modification can be constructed. There are exceptions to this requirement for replacement units or maintenance, but if there will be an increase in emissions, a permit usually will be needed.

PURPOSE OF PERMITTING PROGRAM

- Site new or modified sources without causing a new violation of air quality standards, or exacerbating an existing one
- Provide for systematic review of proposed construction and operation of sources to ensure compliance with air quality regulations

---

Seasonal sources sometimes do not require permits, depending upon the permitting agency's rules. Temporary sources also often do not require permits, provided the source only operates temporarily within a time specified in the agency rules. Usually, a source is considered temporary if it operates for fewer than 180 days, although agency rules vary when the 180 days are not consecutive.

In the past year, stringency has increased for portable or temporary sources, and these sources must meet registration requirements. Once a source is registered, it can operate in other jurisdictions that accept the same registration requirements. Consultation with the agency may be needed to determine whether a permit or registration is required for portable or temporary sources.

The recent trend in permitting is to issue a permit for an entire process or plant. Such permits allow more flexibility in the operation of a source than permits for individual pieces of equipment. Facility permits are required for federal Title V permitting, discussed below.

Mobile and area sources usually do not require permits. Mobile sources emissions are managed through the mobile source provisions of the Clean Air Act, which limit emissions from motor vehicles. Under recent rulemaking, EPA also is responsible for emissions of non-road engines manufactured after June 19, 1994.[2] Motor vehicle engine emission limits have become progressively more stringent, and mobile sources are becoming a larger fraction of the total emissions in urban areas.

Area sources such as gas stations, dry cleaners, and residential heating systems, also do not usually require permits. However, the size of a source requiring a permit has been decreasing in nonattainment areas, and some types of area sources are now required to obtain air permits.

REGULATORY REQUIREMENTS DIFFER DUE TO

- Jurisdiction (which state or region within a state)
- Attainment status of area for pollutants being emitted
- Source type, size, and other characteristics

In the section below, thresholds for permitting are discussed. Sources that do not usually require permits may need permits if their emissions exceed certain thresholds.

A. Thresholds for Permitting

Several important thresholds for air quality permitting are shown below. They are important because they are bright lines that affect the kind of permit you must obtain, and the requirements that must be met to obtain a permit. The thresholds associated with each of these requirements are most often based on emissions or increases in emissions. In other cases, they are based on air quality impacts. The specific thresholds for nonattainment new source review (NSR) and PSD are discussed as part of the descriptions for those programs.

THERE ARE SEVERAL THRESHOLDS FOR PERMITTING

- When a permit is required
- When BACT is required
- When offsets are required
- When NSR is required
- When PSD is required

---

1. Permit Thresholds

In most jurisdictions, there is a general rule that all sources of air pollutant emissions and all emission control devices must have air quality permits. However, this very inclusive rule is mitigated by exclusions to the permitting requirement. A permitting agency will list several kinds of sources that do not require permits. Usually, there will be a threshold on emission rate. For example, in the Bay Area Air Quality Management District (CA), a source emitting greater than 10 lb/day of criteria pollutants on its highest day must obtain an air quality permit.[3] This threshold is not the same in every state or local region.

Another category of exclusions is for particular types of sources. One common exclusion is for heating systems for single-family dwellings.[4] In some jurisdictions, this exclusion extends to dwellings for up to four families.[5] In that instance, an air permit would not be required for an apartment house with four apartments, but a permit would be required for a building with five apartments unless another exclusion applied.

A second common exclusion is for boilers with heat input less than 1 million Btu/hr.[6] If a state or region has both this exclusion and the four-family dwelling exclusion, a building might have five apartments, but if its heating system is smaller than 1 million Btu/hr, it would still not require a permit.

If it appears, after you have reviewed the applicable regulations, that your air pollution source may qualify for an exclusion, you should call your permitting agency and ask. Agencies commonly receive these kind of inquiries, and staff are usually very familiar with the limitations on excluded sources. If you receive an opinion from your phone call that your source is indeed excluded from permit requirements, you should write a letter to the agency summarizing your phone contact (with the name of the person you talked to and the date). The agency is not likely to respond to your letter, but if an issue arises later, this documentation, and proof that it was furnished to the agency, provides a powerful argument that you acted in good faith.

2. Potential to Emit

Emission thresholds identified by permitting agencies are almost always based upon the *potential to emit* of the source or facility. The potential to emit is the maximum emissions a source could release operating at full capacity full time — 24 hr/day, 365 days/year, unless there are enforceable limits preventing this. For some types of sources, potential to emit is easily calculated. For a boiler or other heating device, the potential to emit would be based on the design or nameplate capacity of the device and 8760 hr/year of operation. For other kinds of sources, determining potential to emit is not so straightforward. For a paint spray booth, a hypothetical potential to emit would be with all spray guns operating all the time. If a spray gun had the capacity to spray 5 gallons of paint per hour, that would amount to a potential to emit of 131 tons/year of VOC.* Such an estimate is ridiculous, since it does not consider the limitations of supplying the paint gun with a 55-gallon drum of paint every 2 days, moving parts to be painted into and out of the booth, cleaning the gun periodically, having personnel take mandated breaks and lunch. For painting operations, and other near-infinite emitters, conditions on operations are usually used to limit the potential to emit.

---

* Assuming 6 lb/gal of VOC in the paint and operation 8760 hr/year.

Paperwork, fees, inspections, and other requirements must be accepted as corollary to having an air quality permit. If normal operations of a facility do not exceed permit thresholds, accepting constraints on operations may be necessary to avoid permitting. This option has been selected by several facilities as a means to avoid federal Title V permitting, as discussed below.

Physical constraints that limit emissions may also be a basis for avoiding permitting. For example, for the paint booth example given above, an automotive body shop demonstrated that the maximum rate of painting was limited by the number of cars that could be moved into and out of the booth in a workday, and the amount of paint required to paint a car.

3. BACT Threshold

The next threshold that is usually considered specifies the emission levels that require the source to utilize best available control technology (BACT) to control or limit emissions. The methods for determining what constitutes BACT for a particular source are discussed in Chapter 3. The threshold for requiring BACT is different for different pollutants, so in order to determine whether BACT is required, emissions for all pollutants from a source must be estimated. Methods for estimating emissions are discussed in Chapter 2. If emissions from a source are just slightly above the BACT threshold, utilization of BACT will result in the emissions after control being below that threshold. For sources in that situation, it would be advisable to include control measures in the original design so that the resulting emissions are below the BACT threshold.

At the present time, rules in most jurisdictions require BACT on all sources that must have a permit, so the above strategy is not helpful. However, some jurisdictions have not updated their rules, and still have fairly high thresholds for BACT (150 lb/day used to be a common threshold), so checking the rules can be important in trying to limit control equipment expenditures.

4. Offset Thresholds

Unlike most current BACT thresholds that are the same as permitting thresholds, requirements for emission reduction offsets are usually well above permitting thresholds, and offset requirements are limited to large sources. Historically, offset requirements have been limited to permitting in nonattainment areas, but recently offsets have also been required in attainment areas.[7] Obtaining offsets is discussed in Chapter 4.

III. NEW SOURCE REVIEW IN NONATTAINMENT AREAS

New Source Review (NSR) is a permitting program established under the Clean Air Act to require and administer permits for new and modified sources of air pollutant emissions in both attainment and nonattainment areas. In attainment areas, NSR is the PSD program. This is discussed in the next section. In nonattainment areas, NSR is used. In the subsections below, the applicability of the NSR program is discussed first, then requirements that must be met to obtain a permit are presented.

NEW SOURCE REVIEW INCLUDES

- Permitting new major sources in nonattainment areas under NSR
- Permitting new major sources in attainment areas under PSD
- Permitting major modifications of sources in both attainment and nonattainment areas

---

NSR applies to both new and modified sources. Minor sources that must undergo NSR are included because they have been identified for inclusion in a SIP. New sources in nonattainment areas are considered major sources subject to NSR if emissions exceed the thresholds shown in Table 1-2. Major modifications to major sources must undergo NSR if emissions exceed major modification thresholds shown in Table 1-3.

Table 1-2   New Source Review Thresholds for Nonattainment Areas

Pollutant	Area Designation	Threshold (tpy)
Ozone precursors (NO_x, VOC)	Marginal, moderate Serious Severe Extreme	100 50 25 10
Inhalable particulate matter (PM_{10}) and PM_{10} precursors (NO_x, SO_2, VOC)	Moderate Serious	100 70
Carbon monoxide Nitrogen oxides Sulfur dioxide	Any nonattainment area	100

Source: 40 CFR 51.165

Requirements for NSR in nonattainment areas are intended to ensure that requirements in permitting new sources will result in reasonable further progress toward the goal of reaching attainment. To ensure this, NSR requires that emissions from new sources are at the Lowest Achievable Emission Rates (LAER) and that there is no net increase in emissions.

LAER is defined in Chapter 3. LAER is at least as stringent as, and maybe a more stringent emission control technology than, BACT, since it is not constrained by any economic criterion.

No net increase in emissions is achieved through requirements for emission offsets, as discussed in Chapter 4. NSR sources must locate and reduce emissions from existing sources to offset the emission increases from the new or modified source. The reductions from existing sources must be greater than the new source emissions to ensure the further progress identified above. The offset ratio increases with the severity of the nonattainment area.

Table 1-3   Major Modification Thresholds for Nonattainment Areas

Pollutant	Area Designation	Threshold (tpy)
Ozone Precursors (NO_x, VOC)	Marginal, moderate	40
	Serious, severe	25
	Extreme	0
PM_{10}	Moderate, serious	15
CO		100
NO_x		40
SO_2		40
Pb		0.6
Fluorides	Any nonattainment area	3
Sulfuric acid mist		7
H_2S		10
Total reduced sulfur		10
Reduced sulfur compounds		10

Source: 40 CFR 51.165

NONATTAINMENT REQUIREMENTS APPLICABLE TO MAJOR SOURCES AND MAJOR MODIFICATIONS

- Lowest achievable emission rate (LAER)
- Offsets or demonstration of net air quality benefit
- Statewide compliance for all sources owned by applicant
- Alternative analysis

There are three additional requirements under NSR in nonattainment areas. The first is that other major sources of air pollutants under the same ownership in the state must be in compliance with all applicable air quality regulations. This requirement is sometimes difficult to interpret for large corporations with several divisions, or for government owned sources such as military bases; but in most cases it has been interpreted narrowly, affecting only a single division of a company, or bases under separate base commanders.

The second additional requirement is that an alternative analysis must be carried out to demonstrate that the siting of an air emission source within a nonattainment area has greater benefits than the environmental costs associated with the project. This analysis must include consideration of secondary growth that could occur in the impact area as a result of the new source as well as non-air quality costs and benefits. The alternatives analysis may require analysis under the national environmental policy act (NEPA) or a state environmental policy act (SEPA).

The third additional requirement is that sources that may impact visibility in mandatory Class I areas* must be reviewed by the Federal Land Manager for that Class I area. This review can be quite extensive, and often requires air quality modeling (see Chapter 5) to determine whether visibility standards can be met.

---

* Areas considered pristine such as national parks. Class I areas are discussed further under PSD.

IV. PREVENTION OF SIGNIFICANT DETERIORATION PERMITTING

The federal prevention of significant deterioration (PSD) program was established in 1978 as a result of a lawsuit that argued that the Clean Air Act Amendments of 1977 required that a program be established to prevent degradation of air quality in regions of the country that were currently in attainment.[8] The PSD program requires that major sources in attainment areas obtain permits, and that emissions from these sources cannot cause deterioration of ambient air quality beyond certain increments — and in no case, beyond the ambient air quality standards.

PREVENTION OF SIGNIFICANT DETERIORATION (PSD) GOALS

- Ensure economic growth will be in harmony with preservation of existing clean air areas
- Protect public health and welfare from any adverse effects in attainment areas
- Preserve, protect, and enhance air quality in areas of special natural, recreational, scenic, or historic values

---

A. Major Sources Subject to PSD

The PSD program is only applicable to *major* sources of regulated air pollutants other than HAPs in attainment areas. *Major* sources are defined by their annual emission rate and the industrial category of the facility. A facility is major if its emissions of any regulated pollutant exceed 100 tons/year and it is in one of the 28 industrial categories shown in Table 1-4. If the facility is not in one of these 28 listed categories, it must have emissions above 250 tons/year to be regulated as a major PSD source.

MAJOR PSD SOURCE DEFINITION

- Any source listed on Table 1-4 with a potential to emit greater than 100 tons/year of any regulated pollutant except HAPs
- Any other source with potential to emit greater than 250 tons/year

---

Facilities must keep track of total emissions to determine whether they become a major source with the addition of new equipment or change in method of operation. Sources that change from one permitted fuel to another, or increase hours of operation within the provisions of an existing permit, are not newly subject to PSD. However, if they were permitted for a fuel with higher potential to emit than a previously used fuel, and they were permitted for more hours of operation than previously used, the facility potential to emit was already above the major source threshold. If an existing facility not subject to PSD permitting

Table 1-4 Major Stationary Sources Under PSD if Emissions >100 Tons/Year

Fossil fuel-fired steam electric plants greater than 250 million Btu/hr heat input
Coal cleaning plants (with thermal dryers)
Kraft pulp mills
Portland cement plants
Primary zinc smelters
Iron and steel mill plants
Primary aluminum ore reduction plants
Primary copper smelters
Municipal incinerators capable of charging more than 250 tons of refuse per day
Hydrofluoric, sulfuric, and nitric acid plants
Petroleum refineries
Lime plants
Phosphate rock processing plants
Coke oven batteries
Sulfur recovery plants
Carbon black plants (furnace process)
Primary lead smelters
Fuel conversion plants
Sintering plants
Secondary metal production plants
Chemical process plants
Fossil fuel boilers (or combinations thereof) totaling more than 250 million Btu/hr heat input
Petroleum storage and transfer units with total storage capacity exceeding 300,000 barrels
Taconite ore processing plants
Glass fiber processing plants
Charcoal production plants

Source: 40CFR 52.21(b)(1)(i)(a)

is modified so that the total facility potential to emit exceeds the PSD threshold (100 TPY or 250 TPY depending on the industrial source category), then the facility becomes a major source and must obtain a PSD permit for any future major modification (see below).

B. Pollutants Subject to PSD Permitting

Under federal law, sources emitting any regulated pollutant or its precursors except hazardous air pollutants (HAPs) are subject to permitting under the PSD program.* However, many states and local regions that have assumed delegated authority under the PSD program apply PSD permitting requirements to sources of other pollutants as well.** In addition, significant emission rates are identified

---

* Including total suspended particulates, which is no longer a criteria pollutant, but which still has standards set under some new source performance standards (NSPS).
** This is the result of the history of the PSD program. Prior to the 1990 CAAA, there were 14 pollutants subject to PSD permitting. These were the criteria pollutants and eight others: asbestos, beryllium, mercury, vinyl chloride, fluorides, sulfuric acid mist, hydrogen sulfide, and total reduced sulfur compounds (including hydrogen sulfide). When Title III of the 1990 CAAA excluded HAPs from the PSD program, four of these eight pollutants were dropped because they are HAPs (see Chapter 6). Many states and local regions retained these HAPs in their PSD programs

for only ten pollutants, discussed below under Major Modifications to PSD Major Sources.

Since there are only PSD increments for three criteria pollutants, PSD permitting does not require increment consumption analysis for the other pollutants subject to the program. Nonetheless, sources of these pollutants must still meet the other PSD permitting requirements.

C. Major Modifications to PSD Major Sources

Once a facility has a PSD permit, it must obtain a revision to that permit when it increases its emissions above the major modification threshold, shown in Table 1-5. The increase in emissions to be compared to the threshold is the net increase, with emissions from new emission units decreased by any reduction due to shutting down or decreasing emissions from existing emission units.

**Table 1-5   Thresholds for Major Modifications to
PSD Major Sources**

Pollutant	Emissions (tons/year)
VOC	40
CO	100
NO_x	40
SO_2	40
PM_{10}	15
Lead	0.6
Fluorides	3
Sulfuric acid mist	7
H_2S	10
Total reduced sulfur	10
Reduced sulfur compounds	10

Source: 40CFR 52.21(b)(23) as revised by Clean Air
Act §112(b)(6)

Netting the differences in emissions between new emission units and emission reductions requires that you count the potential to emit of the new emission unit, constrained by any conditions acceptable to the applicant, minus the *actual* emissions of the unit being shut down or reduced in operation, averaged over the previous 2 years. Thus, although the applicant may be replacing one boiler with another, there may be a net increase in emissions if the new boiler operating schedule includes more hours than the shutdown boiler. This is often the case in a multiple boiler plant where the boiler being shut down was the oldest, least efficient unit. On the other hand, the new boiler will probably have lower emissions both because of new technologies and the requirements of new source performance standards (NSPS; see below) which apply regardless of whether PSD permit review is required. The conditions governing emission reductions for netting under PSD are consistent with those for emission reduction credits used as offsets, as discussed in Chapter 4.

The details of determining whether a modification must undergo PSD review can get complicated for a source with multiple emitting units constructed at different times. Several small modifications can occur if they are unrelated, i.e., not part of the same process. If they are part of the same process, the sum of the multiple emitting units' emissions must be compared to the PSD modification threshold. Once also must count *debottlenecking,* or increases in emissions of existing units if the new construction enabled otherwise impossible increases in production.

After a source is a major PSD source, increases in emissions of any PSD pollutants must be compared to the modification threshold, not just the pollutant(s) that caused the source to originally be designated as major.

D. PSD Permit Requirements

Once a source has been determined to require a PSD permit, it must comply with requirements for Best Available Control Technology (BACT), and analyses of ambient air quality impacts, Class I impacts, and air quality related values impacts must be conducted.

PSD REQUIREMENTS

- Best Available Control Technology
- Analysis of impacts on NAAQS and PSD increments
- Class I analysis
- Air Quality Related Values (AQRV) analysis

---

1. BACT Analysis

To obtain a PSD permit, all emission units at the proposed source facility must utilize BACT for control of air pollutant emissions. The process for determining what technology constitutes BACT for a particular category of equipment is described in Chapter 3.

2. Air Quality Impact Analysis

In a PSD permit application, you must include an air quality impact analysis (AQIA). The purpose of this analysis is to determine whether the ambient pollutant concentrations due to the proposed source will either cause an exceedence of the NAAQS or of PSD increments.

a. NAAQS Determination. To determine whether the project emissions will cause an exceedence of the NAAQS, two pieces of information are needed. First, it is necessary to know what the current background concentrations of PSD pollutants are in the vicinity of the proposed project or modification. This is determined either from ambient air quality data obtained using existing monitors in the vicinity of the proposed project, or through preconstruction monitoring at a station set up for that purpose. The permitting agency decides whether available data are adequate to determine background concentrations. If the predicted air quality impacts of the proposed project are below the thresholds shown in Table 1-6, no preconstruction monitoring is required. If the predicted impacts exceed these thresholds for PSD pollutants, and the permitting agency determines that available data are not adequate for determining background concentrations in the vicinity of the proposed project, a minimum of 1 year of preconstruction monitoring will be required. This background monitoring must be complete before the PSD permit application is submitted, so if required, preconstruction monitoring can add a year to the time required to obtain a PSD permit.

Table 1-6   Predicted Concentration Ceilings to Avoid Preconstruction Monitoring

Pollutant	Concentration ($\mu g/m^3$)	Averaging time
CO	575	8 hr
NO_2	14	Annual
PM_{10}	10	24 hr
SO_2	13	24 hr
Ozone	100 tpy VOC	
Lead	0.1	3 month
Fluorides	0.25	24 hr
Total reduced sulfur	10	1 hr
H_2S	0.2	1 hr
Reduced sulfur compounds	10	1 hr

Source: 40CFR 52.21(h)(8)(i)

The second information needed is the predicted impact of the proposed project or modification emissions. This is obtained using dispersion modeling. Approaches to dispersion modeling are discussed in Chapter 5. The predicted impacts of the project emissions are added to the background concentrations to determine whether any NAAQS have been exceeded. If a NAAQS is exceeded due to the proposed emission increases, the permit will be denied. If modeling shows a predicted concentration exceeding the NAAQS, the applicant should review the modeling and the emissions estimates to determine whether there are measures that can be taken to reduce the predicted impacts. These would include more sophisticated modeling or reductions in project emissions through the use of more efficient control technologies or other emission reduction measures.

b. PSD Increment Exceedence Determination.   PSD increments are allowable increases in concentrations of specific pollutants in attainment areas due to new sources of emissions. They were established for three classes of attainment areas, as shown in Table 1-7. The smallest increments are for Class I areas. Class I increments allow the least degradation of air quality. All national parks, monuments, and wilderness areas are designated Class I areas. Other portions of the country can petition to become Class I areas, and several Native American reservations have been given Class I designation.

Table 1-7   PSD Increments for New or
             Modified Sources

Pollutant	Class I ($\mu g/m^3$)	Class II ($\mu g/m^3$)
SO_2, annual	2	20
SO_2, 24 hr	5	91
SO_2, 3 hr	25	512
PM_{10}, annual	4	17
PM_{10}, 24 hr	8	30
NO_2, annual	2.5	25

Source: 40CFR 52.21(c)

All remaining attainment areas are designated Class II. PSD increments were also established for Class III areas, allowing greater degradation than for Class II areas, but no area of the country has petitioned to become a Class III area.

Since PSD permits must be obtained before a source is constructed, determination of whether the emissions from a proposed or modified source are within the appropriate increments is done by modeling the emissions of the proposed source. The requirements for PSD modeling are discussed in Chapter 5.

3. Class I Analysis

If the proposed source or modification is within 100 km of a Class I area, concurrence of the Federal Land Manager (FLM) for the Class I area must be obtained before a PSD permit can be issued. To obtain FLM concurrence, the applicant must demonstrate that Class I increments are not exceeded in the potentially affected Class I area(s), that visibility in the Class I area will not be impaired (visibility modeling is discussed in Chapter 5), and that other air quality related values do not deteriorate. The FLM may request mitigation measures to assure that AQRV are maintained.

4. Air Quality Related Values Analysis

For any PSD permit, an evaluation of air quality related values (AQRV) must be conducted. The AQRV analysis includes impacts on soils, vegetation, visibility, and growth of air pollutant emissions from the proposed project. Generally, this evaluation is carried out for all pollutants emitted from the project, not just those subject to PSD. In particular, an analysis of ozone impacts is required when a PSD permit is being sought for attainment pollutants in an ozone nonattainment area.

The impact on soils and vegetation is to be developed for soil and vegetation types in the impact area for the proposed project. Available research data are usually for the region, since vegetation types do not change over areas as small as the impact area of a single project.

The visibility impacts are determined using a visibility model. Visibility modeling is discussed in Chapter 5.

A growth impact analysis must first estimate the increases in population and commercial and industrial activities likely to result from the proposed project. Then air pollutant emissions associated with those increased activities must be estimated and potential impacts on air quality predicted using computer dispersion modeling.

V. PROHIBITIONARY RULES

The emissions of a proposed source or emission unit will be limited by several requirements. As identified above and discussed in Chapter 3, sources are required to utilize BACT to limit emissions. The allowed PSD increments will effectively limit source emissions. Other emission limits include NSPS and other prohibitionary rules.

A. New Source Performance Standards

NSPS are maximum allowable emissions for new or modified sources.[9] The applicability of each NSPS is specified in the NSPS. NSPS have been developed for more than 60 categories of emission sources[10] and represent the best system of emission reduction that has been adequately demonstrated, with consideration of emission reduction potential, cost, energy, and environmental impacts. The NSPS only apply to sources constructed after each standard was promulgated. NSPS have been modified as new control technologies become available, so for several of the NSPS, there are ranges of applicable dates, and a newer standard applies for more recently constructed sources.

The statutory basis for NSPS sounds like the basis for BACT, but BACT levels are often much lower than those under NSPS. In fact, the NSPS are the statutory floor for BACT decisions. BACT levels are probably lower because BACT is determined on a case-by-case basis, whereas NSPS levels are negotiated as industrywide standards in a regulatory process that usually takes several years.

B. Control Technology Guidelines

To address technology requirements for nonattainment, EPA has adopted Control Technology Guidelines (CTG) to be used by regulatory agencies in establishing control technologies for existing sources.[11] Each set of CTG was developed to establish Reasonably Available Control Technologies (RACT) for a specific set of sources or source categories. The CTG also consider cost, energy, and environmental impacts, but since they are oriented to existing sources, they are generally not as stringent as NSPS. Since the CTG are not regulations, they are not included in the *Federal Register* or *Code of Federal Regulations (CFR)*.

VI. TITLE V OPERATING PERMITS

Title V of the CAAA established a federal operating permits program. The program is administered by the states or local regions under rules established by those agencies and approved by EPA as being as stringent as the federal program.[12] Most states are currently implementing programs that have received interim approval by EPA. This means that they can issue valid Title V permits to applicable sources in their jurisdiction for 2 years after the agency has been granted interim status. Final Title V authority will be issued by EPA after a program that has received interim approval corrects program deficiencies perceived by EPA to need correction so that the stringency of the federal program is achieved, but to be not so serious that permits cannot be issued. The program changes needed are specified in the *Federal Register* notice that issues interim approval for each agency.

A. Applicability

Title V permits must be obtained by all major sources of regulated air pollutants including major sources of hazardous air pollutants (HAPs).* The thresholds for a source being major depend upon the attainment status of the area in which the source is located. These thresholds for attainment areas and nonattainment areas of increasing severity are shown in Table 1-8.

The applicability threshold is based upon the potential to emit a source, although if a source can document actual emissions below half of the applicable potential to emit threshold, EPA guidance is that a Title V permit is not necessary.[13]

WHO NEEDS A TITLE V PERMIT?

- Major sources (Table 1-8)
- Sources subject to NSPS
- Major HAP sources
- PSD/NSR sources
- Acid rain sources (subject to Title IV)
- Municipal waste incinerators

The intent of the Title V program is to permit major sources, but a facility may avoid Title V permitting by limiting its potential to emit through a federally enforceable state or local region *synthetic minor* permit. The *synthetic minor* permit specifies conditions of operation of an otherwise major source so that its emissions cannot exceed the Title V thresholds. The conditions limiting potential to emit must be enforceable so that an agency inspector can determine whether they are being met. It is likely that record keeping will be necessary to provide documentation that these conditions are being met.

Estimating emissions for a source is subject to several definitions. Fugitive emissions are not included in the emission estimate to determine applicability unless the source is in one of the 28 categories identified in the PSD program

---

* Regulated pollutants include the criteria pollutants, ozone depleting substances regulated under CAAA Title VI, pollutants regulated under NSPS or other Clean Air Act regulations, and HAPs. Pollutants identified in the Section 112(r) program for accidental release prevention are not included in this definition.

Table 1-8 Applicability Thresholds for Major Stationary Sources Requiring Title V Permitting (Emissions in Tons per Year)

Area classification	Oxides of nitrogen (NOx)	Organic compounds (VOC)	Particulate matter (PM10)	Carbon monoxide (CO)	Individual toxics	Total toxics	Other regulated pollutants
Attainment	100	100	100	100	10	25	100
Marginal	100	100	100	100	10	25	100
Moderate	100	100	100	100	10	25	100
Serious	50	50	70	100/50*	10	25	100
Severe	25	25			10	25	100
Extreme	10	10			10	25	100

* Serious nonattainment areas for carbon monoxide (CO) in which stationary sources contribute significantly to CO levels shall submit a plan which provide that major stationary sources include any source with CO emissions greater than 50 TPY [CAA Section 187(c)(1)].

Source: Clean Air Act Amendments of 1990, Title I

for 100 TPY thresholds, shown in Table 1-4. Fugitive emissions must also be included for sources of HAPs. Emissions from "trivial sources" do not need to be included.[14] Emissions from sources identified in a state or local region program that are "categorically exempt" also do not need to be included if the program has final approval from EPA.

However, emissions from all presently permitted sources, and other sources not trivial or categorically exempt, must be included in the emission summary to determine whether the source is major, and therefore must obtain a Title V permit.

B. Application Preparation

Preparation of the Title V application for a major facility is a major undertaking. Each of the states or local regions with Title V permitting authority is required to develop application forms for use by applicants. Most agencies insist on the use of their forms. Because of the substantial burden of review placed on the agencies, its a good idea to have all the applications organized the same way so that the reviewer can conduct his or her review efficiently. For several of the states, this was the first time there had been a comprehensive permit program, so there were not already standard forms to use.

The challenge faced by the applicant with new forms for a new program could be minimized with some of the approaches described below. These include:

- *Consistency.* Within the form, the same information must be put down more than once. Often, in fact, several times. There must be consistency among these entries. For example, if you state that the facility operates 8 hr/day, emission estimates cannot be based on a 16-hr day. Consistency requires careful review, especially if several people are working on different parts of the application.
- *No new numbering systems.* It can be an efficient abbreviation to number emission units or other devices and make subsequent references by number rather than by the name of a unit. However, plant personnel are already familiar with descriptions of devices, such as the "West paint booth," and it will require extra effort for them to figure out which booth you are talking about. If there are already numbers, such as individual permit numbers for these devices, these can be used, but creating a new numbering system solely for the Title V permit application can create unnecessary confusion. Where possible, use the same numbers already used by the plant. If a numbering system must be created, make it spatially consistent and provide a plant layout to show what device you are referring to with a new number.
- *Guide to the reader.* Annotation in the application can be very helpful to the agency reviewer. If information that might be needed to review a particular page is not nearby, a note indicating where it can be found can reduce time and frustration.
- *Word process.* Wherever possible, word processing should be used to complete the forms. For numerical tables, use spreadsheets. They are much easier to read than handwritten information, and convey to the agency a greater level of professionalism in preparation than pen and ink. Most of the agencies have used word processors to create their forms, and are willing to provide electronic copies for use by applicants. Another benefit of word processing forms is that

word processing search capabilities can be used to locate information in the application.

- *Loose-leaf.* Use loose-leaf notebooks to bind applications. Changes to the application after submission are common, in part because it is a first submission under the Title V program, and loose-leaf binding eases the correction process.
- *Number pages.* The easiest to use numbering system for the reader is sequential from front to back. For the preparer, numbering by sections is easier, since revisions during the preparation do not require renumbering of the whole document. If numbering by sections is used, limit the number of sections and make them easy to find by using tabs.
- *Documentation of estimates.* It is vital that the reviewer be provided with enough information to check the applicant's emission estimates. All assumptions should be clearly stated, and bases for assumptions provided.

C. Application Content

The content of Title V applications was specified in the CAAA. Since implementation of the Title V program, there has been clarification of the content requirements. The application forms provided by the permitting agency are intended to request all required information. The list below should be consistent with information requested.

The application must:

- identify the facility, provide an accurate location (often requested in UTM coordinates), provide a description of processes and products, and provide the Standard Industrial Code (SIC)[15] for the facility;
- estimate the emissions from emission units within the facility;
- identify regulatory requirements applicable to the facility and its emission units;
- identify compliance demonstration methods to be used by the facility to demonstrate compliance with all applicable requirements;
- certify compliance with currently applicable requirements and intent to comply with future requirements; and
- include a certification by a responsible official of the truth, accuracy, and completeness of the application.

Each of these elements is discussed below.

1. Identify the Facility

The facility's name, location, and mailing address should be provided. Since Title V major facilities are defined as contiguous, under common ownership, and with the same two-digit SIC code, the SIC code of the facility is needed. This general portion of the application often also asks for hours of operation and a technical contact person. It would be unusual for this contact person to be the same as the responsible official (see below), but often in correspondence regarding the application, the permitting agency will direct correspondence to the responsible official. Make sure that the responsible official provides copies of such correspondence to the technical contact person so that any needed response can be prepared.

2. Estimate Emissions

For regulated emission units, the potential to emit as limited by any applicable regulations must be provided. Estimates must also be provided for other units if this information is needed — for example, to determine permit fees. Methods for estimating emissions are given in Chapter 2. It is not necessary to conduct source tests or other measurements not currently taken to make emission estimates for the application. However, source testing or other measurements may be necessary to demonstrate compliance after the permit has been issued.

3. Identify Regulatory Requirements

A primary intent of the Title V permitting program is to identify all of the regulatory requirements applicable to a facility and its emission units, and to specify methods that will be used to determine whether the facility is in compliance with those requirements. It is important that all potentially applicable regulations be included. If a regulation is specified in the application as not applicable to the facility, enforcement cannot be instigated against the facility regarding that regulation during the time the application is being reviewed. If this nonapplicability is included in the permit, under the permit shield, no enforcement is allowed. Only regulations included in the permit as applicable to the facility are applicable. Other existing regulations are not.

The regulations applicable to the facility as a whole, or to all emission units within the facility, should be specified first, and then regulations applicable to each emission unit. Only the citation to the regulation is needed.

Conditions contained in existing state or local permits should also be included, provided they are based on promulgated regulations. In some preexisting permits, conditions were included that were agreed to by the applicant at the time the permit was issued, but are not needed to comply with applicable regulations. These conditions are not required to be part of a Title V permit.

Regulations or conditions that cannot be enforced are also not required to be part of a Title V permit. An example of this kind of regulation would be an ambient air quality regulation. Each of the states has regulations requiring that areas of the state be in compliance with ambient air quality standards, but this regulation cannot be enforced against an individual facility. Emission limitations can be enforced, but the ambient air quality standards cannot.

4. Identify Compliance Demonstration Methods

For each regulatory requirement applicable to the facility or any of its emission units, a method must be identified for the facility to demonstrate that it is in compliance with that requirement for the duration of the permit. For example, if there is a requirement to maintain emissions of a boiler below specified levels, the applicant may propose a source test to demonstrate that emissions meet the requirement, and record keeping on fuel use by the boiler. The fuel use can be correlated with emissions by using an emission factor developed from the source test, so the fuel use becomes a surrogate for boiler emissions. Documentation that fuel use is below the rate that results in emissions beyond allowable limits demonstrates that the boiler is remaining in compliance.

Record keeping will be an important part of each Title V permit, because records are needed to demonstrate continued compliance. Even if a continuous emission monitoring (CEM) system is in place, records of the emissions measured are needed to demonstrate compliance.

Reporting of compliance demonstrations is required every 6 months. This self-reporting is intended to document any noncompliance, and enforcement action may be taken on any noncompliance reported.

A statement is also required as part of the application that the facility will comply with future requirements when they are promulgated.

5. Certification by Responsible Official

A responsible official of the facility must sign the application. The official must attest to having made a diligent effort at assuring that the application is true, complete, and accurate based on reasonable inquiry.

The responsible official who signs the permit application must recognize that he or she is attesting to a considerable amount of information about the facility being permitted, including intent to comply with future requirements. For a corporation, the responsible official must be a corporate officer or other person in charge of a principal business function. It can be a plant manager if the facility has at least 250 persons employed or $25 million in sales or expenditures, and if the plant manager has been delegated authority to sign as responsible official in advance. For a partnership, the responsible official is a general partner. For a sole proprietorship, it must be the proprietor. For a local or state government agency, it must be the principal executive officer or the ranking elected official. For a military installation, it must be the commanding officer.

Since there are both civil and criminal penalties for violations of Clean Air Act provisions (see Chapter 8), the responsible official must be confident that items attested to in the application are correct.

6. Permit Application Shield

Title V includes a provision for an application shield to protect sources from penalties due to untimely review and approval of their applications. The shield applies only to complete applications and does not start until the application is declared complete. It protects the source from notices of violation due to operating without a permit when the issuance of a permit is delayed. Sources using the shield are still obliged to comply with all requirements in any preexisting permit, even though that permit may have expired. The shield is lost if the source fails to provide timely updates to source information.

7. Other Information

a. Alternate Operating Scenarios.   A provision in Title V allows a facility to specify alternate operating scenarios under which the facility may operate. Alternate operating scenarios are distinguished by different applicable regulations and/or different pollutants emitted. If alternate operating scenarios are included in the Title V permit, the facility operator must notify the permitting agency within 5 days after a different scenario has been implemented.

In practice, there are very few situations that require alternate operating scenarios. It is in a plant's best interest to demonstrate compliance with all applicable regulations, even those that only apply part of the time, and it is also a good idea to include in your Title V permit estimates of the potential to emit for any pollutants likely to be emitted. In this way, the plant maximizes flexibility in its operations. Identifying alternate operating scenarios is one more thing to keep track of, and if the single operating scenario of the plant is written broadly, it can allow all of the flexibility provided by an alternate operating scenario.

b. Noncompliance.   If sources are not in compliance with current requirements when they make application for a Title V permit, the application must include a narrative description of how compliance will be achieved, a detailed schedule of steps that will be taken to move the facility toward compliance (including remedial measures to deal with any consequential impacts of noncompliance), and a schedule for submission of progress reports documenting the actions taken in compliance with the plan. This compliance plan should have the level of detail and specific actions such as those in a consent decree.

A source already should have disclosed noncompliance and developed this compliance plan with the applicable regulatory agency. In many cases it will be a consent decree. If there has not been prior disclosure, the source should expect enforcement action from the regulatory agency or a citizen suit.

c. Acid Rain Program Compliance.   Sources that have received a Phase I permit under the Title IV Acid Rain Program requirements may reference that permit in their Title V permit application for informational purposes, but that permit is separately issued by EPA and only enforceable by EPA. Phase II acid rain permits may be a part of Title V permits, and must be a part of Title V permits after Phase II permit renewal. The mechanism for combining the Title IV and Title V permits is at the discretion of the state or local regional permitting authority.[16]

D. Title V Permit Content

Title V permits will contain some important provisions not often found previously in air quality permits. Each of these is discussed below.

The permit will contain emission limitations and standards that must be met by the facility. The emission limits will be based on the permit application and any other constraints imposed by NSPS, MACT, and prohibitionary rules established by the regulating agency.

The standards with which the facility must comply for the duration of the permit must also be specifically included in the permit. Under the provisions of Title V, only the permit conditions will govern the permitted facility. If more stringent rules are passed subsequent to the permit being issued, the facility will need to comply on the schedule identified in the new rules, but the only existing rules with which the facility must comply are those identified in the permit.

The permit will be for a fixed term not to exceed 5 years. All the states have adopted a 5-year permit term, because the administrative effort required for permit renewal will be increased if the term is shorter. When a permit comes up for renewal, any change in the SIP can be imposed as permit conditions, including requirements for reduced emissions or more stringent controls.[17]

Monitoring, record keeping, and reporting requirements provide the means for the regulatory agency to ensure that the permitted facility is continuing to comply with emission limitations. Requirements such as daily record keeping of volatile organic compound use and continuous emission monitoring (CEM) for stacks are possible requirements, but the applicant must propose compliance methods, which may be accepted or changed by the permitting agency.[18]

The permit will include specification of any allowances that are available under the Title IV acid rain program.

The permit will include a severability clause that provides that any provision of the permit found to be invalid in a court action is separate and distinct (severable) from the remaining provisions and does not affect the remainder of the permit.

Several "standard" conditions will be included in the permit. It can be reopened for cause, such as evidence of false representations in the original permit application. It conveys no property rights, that is, obtaining a permit does not convey title to the equipment to the permit holder. Information requested by the permitting agency must be provided. Permit fees must be paid in order for the permit to remain valid.

The permit will contain several compliance provisions, such as the right to inspection and entry by regulatory agency staff or their representatives, access to records by the agency, a schedule for compliance consistent with the compliance plan (if any) provided in the application, and a provision for annual compliance certification, which may include source testing if required by the agency.

E. Permitting Actions

There are several different permitting actions identified in Title V, with varying levels of review and significance for the permitted facility. They include the following:

- Initial permits
- New sources
- Minor modifications
- Significant modifications
- Administrative amendments
- "Other" modifications
- Reopening for cause

1. Initial Permits

The initial permits are those required by Title V for all existing emission sources above the major source threshold described above. These permits will be used by EPA to improve documentation of air emission sources and provide a much improved data base of sources in jurisdictions that have not previously had air permit programs. The timetable on these initial permits has slipped from what was initially stated in the CAAA. That program called for states or local regional jurisdictions to propose a Title V permit program to EPA by November 15, 1993, 3 years after the passage of the Amendments. Those programs had all been submitted by May 15, 1995, and most of them had been approved, although some late submittals were not expecting approval until fall of 1996. Title V permit applications are to be submitted by all major sources in a jurisdiction no later than 1 year after program approval. After a permit application is submitted, the permitting agency has 60 days to declare the application "complete," defined by EPA as having enough information to start processing. The items identified by EPA as necessary to determine an application to be complete include[19]:

- information to the extent needed to determine major source status, to verify compliance with applicable requirements, and to compute a permit fee
- a plan and timetable for achieving compliance; if there is current noncompliance at a facility
- description of the emissions of all regulated pollutants for each significant (i.e., not insignificant) emission unit,
- certification of the truth, accuracy, and completeness of all information submitted, and that the source is in compliance with "all applicable requirements" signed by a responsible official.

2. New Sources

New sources must obtain Title V permits as well as applicable NSR or PSD permits. State or local regional agencies may integrate their Title V and NSR/PSD programs so that one application will address both NSR/PSD and Title V. NSR/PSD permitting is discussed earlier in this chapter.

3. Administrative Amendments

Administrative amendments to permits include correction of typographical errors, changes in the name, address, or phone number of the facility, changes in ownership or operational control, changes to more frequent monitoring or reporting, and NSR modifications provided NSR for the agency is substantially equivalent to Title V. A permitting agency must act on administrative amendments within 90 days, and the facility can make the changes indicated in the amendments immediately after filing the application. The permitting agency must send the revised permit to EPA.

4. "Other" Modifications

"Other" modifications would be changes in emissions of less than 5 to 10 tons per year. The level is determined by the permitting agency. These "other" modifications may be made immediately upon application, and there is no EPA or public review until the permit comes up for renewal.

5. Minor Modifications

Minor modifications to a permit:

- increase emissions by more than 5 to 10 tons per year but less than the threshold levels for Title I,
- do not result in any violation of applicable requirements,
- do not relax monitoring, reporting, or record-keeping requirements,
- do not propose changes that would allow the source to avoid an applicable requirement; do not require a case-by-case determination of an emission limitation or other standard; and
- do not make a modification that would make Title I applicable.

Minor modifications require the permitting agency to notify EPA within 5 days, provide 45 days for EPA review, and require that the request be acted upon within 90 days, or within 15 days after EPA review, whichever is later. There is no public notice for a minor modification, and the source may make changes as soon as the application has been filed. However, compliance with NSR provisions may be required if NSR is triggered by the modification.

6. Significant Modifications

Significant modifications are increases in emissions that exceed threshold levels for Title I NSR, modifications that affect HAP emissions or controls, or modifications that do not qualify as administrative, minor, or "other" modifications. Applications for significant modifications follow the same timetable as applications for initial permits or renewals, and essentially reopen permits for both EPA and public review.

7. Reopening for Cause

A Title V permit may be reopened for cause by the permitting agency or EPA if there are errors in the permit content, revisions are needed to ensure compliance (such as discovery that monitoring or record keeping based on activity data is an inadequate indicator of emissions), or there are new or revised applicable requirements. Reopening is only allowed if the causes for reopening are stated in the permit. If the reopening is for a revised applicable requirement, the permit must be reopened within 18 months after the revised requirement has been promulgated. The timetable and other provisions for reopening a permit are the same as for an initial permit.

F. Fees

Title V includes provisions for fees to cover the costs of this program. The minimum fee is $25 per ton per year, corrected by the Consumer Price Index (CPI) from the date of passage of the Act, November 15, 1990. If a permitting agency proposes a fee of less than $25, it must provide a detailed demonstration of program expenses and income. Fees may also be greater than $25 per ton per year if program costs require it. These fees are only assessed on the first 4000 tons of pollutants.

REFERENCES

1. For a list of regulatory agencies and phone numbers, see *EM,* the Air and Waste Management Associations Magazine for Environmental Managers, October 1996.
2. The attainment status of areas of the country are found in 40 CFR 81, Part C.
3. 40 CFR 89.2.
4. BAAQMD Regulation 2, Rule 2-2-301.1.
5. Indiana Administrative Code Article 2, Rule 2-1-1(a)(5).
6. San Joaquin Valley Unified Air Pollution Control District (CA) Rule 2020, Section 4.2.3.
7. Illinois Rules and Regulations, Title 35: Subtitle B: I-a-201.146 a.
8. Oregon Administrative Rules 340-28-1940(2)(b).
9. Sierra Club v. Ruckleshaus, 412 U.S. 541 (1973).
10. Clean Air Act Section 111.
11. 40 CFR Part 60, Subparts A through UUU.
12. For example, Control Technology Guidelines for Surface Coating of Miscellaneous Metal Parts, EPA-450/2-78/015, Office of Air Quality Planning and Standards, U.S. EPA, 1978.
13. 40 CFR Part 70 specifies the federal Title V program content.
14. Seitz, J. S., "Options for Limiting the Potential to Emit (PTE) of a Stationary Source Under Section 112 and Title V of the Clean Air Act (Act)," Office of Air Quality Planning and Standards, Research Triangle Park, NC, January 5, 1995.
15. Wegman, L. N., "White Paper for Streamlined Development of Part 70 Permit Applications," EPA Office of Air Quality Planning and Standards, Research Triangle Park, NC, July 10, 1995, Appendix A.
16. Standard Industrial Classification Manual — 1987, Executive Office of the President, Office of Management and Budget, U.S. Government Printing Office, PB 87-100012.
17. McLean, B. J., "Guidance on Coordinating Title IV/Title V Permitting Schedules," EPA, April 20, 1995.
18. Clean Air Act, Section 502(b)(5)(C).
19. Clean Air Act, Section 504(b).
20. Wegman, L. N., op. cit.

2   ESTIMATING EMISSIONS

I. INTRODUCTION

Emission estimates are central to air quality permitting. The determination of what regulations apply and what requirements must be satisfied are based on the emissions anticipated from a proposed project. In some cases, a project may not be permittable if the emissions are too high. In most cases, the higher the emissions, the more extensive the requirements that must be satisfied. Emission estimating is usually done in sequential steps, with the first estimates more crude and conservative. If the regulatory requirements based on the crude estimates are acceptable, greater accuracy is not needed. If the first estimates result in costly or otherwise unacceptable requirements, more accurate, less conservative estimates are justified.

In most situations, emissions are only estimated. Since the project is not yet constructed, its emissions cannot yet be measured. Thus the permit application must be based upon estimates. After the project is complete, emissions can be measured using continuous emission monitoring (CEM) or source tests. Source testing is discussed in this chapter, and requirements for verifying emissions are discussed in Chapter 8.

EMISSION ESTIMATING GOALS

- To provide the most accurate estimate of emissions possible given the available data
- Where assumptions must be made, err on the side of public health (health conservative)

In the following sections, approaches to identifying sources of air pollutant emissions are discussed first, and then four techniques for estimating emissions are discussed.

**THERE ARE FOUR ACCEPTED METHODS FOR ESTIMATING
AIR POLLUTANT EMISSIONS**

- Emission factors
- Engineering calculations
- Material balances
- Source testing

There is some overlap among the four emission estimating techniques, but for most purposes, they are distinct. Each technique has particular emission units or situations for which it is most appropriate, and these are discussed along with the methods. Figure 2-1 shows the range of emissions estimating techniques as a function of reliability of estimate and cost.

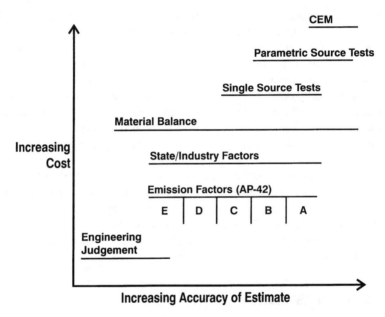

Figure 2-1 Accuracy and cost of different approaches to emission estimation. (Source: US EPA, Compilation of Air Pollutant Emission Factors (AP-42), Fifth Edition, PB 86-142906, U.S. EPA, Research Triangle Park, NC. October 1995.)

II. IDENTIFYING AIR POLLUTANT SOURCES

The first step in estimating emissions from a facility or process is to identify sources that emit pollutants. For a single piece of equipment such as a package boiler, this is clear. However, for a process that includes several pieces of equipment or an entire facility, there may be several sources of air pollutant emissions that are not obvious. All sources must be identified in order to properly permit the facility.

AIR POLLUTANTS CAN BE EMITTED IN SEVERAL WAYS

- Process point sources release pollutants through a stack or vent
- Area sources release pollutants from areas, such as ponds
- Process fugitive sources release pollutants that cannot be collected into a defined release point
- Mobile sources include wheeled equipment, vehicles, and other transportation equipment

---

Different air pollutant source types must be approached differently in order to estimate emissions.

Point sources emit air pollutants from a single identifiable point. This point may be a stack or vent. Stacks are ducts that extend in an upward direction. Vents are horizontal openings. Typical point sources are boilers, furnaces, and other enclosed combustion devices; process vents; paint booths; and particulate control devices such as baghouses, filters, or electrostatic precipitators. Point sources can emit any air pollutant and are only distinguished by having a well-identified location of emissions.

Area sources include other stationary sources that are not point sources. Area sources do not have a well-defined point of emissions, but may be emitting from a line, such as an open channel, or an area such as an evaporation pond.

Fugitive emissions are a category within area sources. Fugitive emissions are releases from leaks in valves, flanges, pumps, and fittings, and although each such leak could be considered a point source, fugitives are usually characterized as area sources because there can be many leaks within a network of piping in a process. Other sources of fugitive emissions may be roads or piles that emit particulate matter and release pollutants over an area rather than from a point.

In some instances, a source that would usually be considered an area source can be treated as a point source because of equipment used to collect emissions and route them to a single release point. Two examples of this would be landfill gas emissions and fugitive emissions from a piping system. Landfill gas emissions are an area source of methane and other organic compounds formed from anaerobic decay of garbage, and which seep up through the soil to the atmosphere. Collection systems of perforated pipes under a slight vacuum can collect these emissions and route them to a single point for control and release. Fugitive emissions from piping systems can be released into a building containing the system, and these emissions become a point source at the point of a building ventilating system vent.

Mobile sources include automobiles, trucks, and other vehicles licensed for public road use, non-road equipment such as construction equipment, and other transportation equipment such as trains, ships, and aircraft. Vehicles licensed for public road use and non-road equipment usually have gasoline or diesel internal combustion engines, but aircraft usually have combustion turbines, and ships may have a variety of prime movers, including diesel engines, boilers, and combustion turbines.

As stringency has increased in air quality regulations in an effort to limit particularly ozone precursor emissions, mobile sources have become subject to permitting requirements. If non-road equipment, such as a forklift, is confined to a facility, its emissions may have to be included in facility emission estimates. Recently, permits or registration also have been required for non-road construction equipment moved from site to site for use.

III. EMISSION FACTORS

Emission factors have become the workhorse of emission estimating techniques. Estimating emissions using emission factors is inexpensive and can provide results that are completely adequate for permitting purposes.

EMISSION FACTORS RELATE MASS EMISSION RATE TO PRODUCTION OR USE RATE

Emission Rate = Emission Factor × Activity Data

$$\left(\frac{\text{Pounds}}{\text{Year}}\right) \quad \left(\frac{\text{Pounds}}{\text{Activity}}\right) \quad \left(\frac{\text{Activity}}{\text{Year}}\right)$$

An emission factor is an estimate of the pollutant emission from an emission unit as a function of a measurable activity associated with the release of that pollutant. It allows you to calculate emissions if you know fuel flow or other similar "activity data" for the unit. Equation 2-1 is used for calculating emissions using an emission factor,

$$E = EF * A \qquad (2\text{-}1)$$

where E = emission, EF = emission factor, in units of emissions per unit of activity, and A = activity level.

As an example, the emission factor for nitrogen oxide emissions from a gas-fired industrial boiler is 140 lb per million standard cubic feet (lb/mmscf) of natural gas burned. Thus an industrial boiler with a natural gas consumption of 40,000 scf/hr (40 mscf/hr) will have NO_x emissions of (Equation 2-2).

$$E_{NO_x} = (EF) \times (A)$$

$$= \left[140 \frac{\text{lb}}{\text{mmscf gas burned}}\right]\left[40 \frac{\text{mscf}}{\text{hr}}\right] \qquad (2\text{-}2)$$

$$= 5.6 \frac{\text{lb}}{\text{hr}} \text{ of } NO_x$$

Emission factors are based upon available source test data. Usually, they do not account for the influence of process parameters other than the activity used as the primary correlation, although often assumptions about other process parameters or characteristics of the emission unit are specified. Thus in the case of natural gas combustion in a boiler, no consideration is given for the design temperature or pressure in the boiler, nor of boiler design characteristics such as quantity of excess air and where it is injected. Because of the simplicity of emission factors, they are easy to use, but may not provide an accurate estimate of the emissions from a process.

An illustration of this is provided in the emission factors for different sizes of gas-fired boilers. The emission factors for NO_x from a natural gas fired boiler change with the size and type of boiler and are shown in Table 2-1. The variation in these emission factors is illustrated in Figure 2-2. There is a 40% increase in the emission rate between boilers less than and greater than 10 million Btu/hr capacity, and a 550% difference between boilers less than and greater than 100 million Btu/hr. Boilers smaller than 100 million Btu/hr are described as industrial boilers, while larger boilers are described as utility boilers. For a facility with a boiler of 150 million Btu/hr capacity in industrial service, the difference between these numbers could be critical in determining the permittability of that boiler.

The difference between these types of boilers is that utility boilers are usually operated to achieve the highest steam temperature possible so as to maximize the efficiency of electricity production. These high temperatures will result in high NO_x emissions. In contrast, industrial boilers often operate unattended, and high steam temperature is not as important as reliability, which is increased with lower combustion temperature and, consequently, lower NO_x emissions. For very small boilers, the lower volume to surface area results in lower combustion temperature and lower NO_x emissions.

As the operating conditions for a process near the edges of those identified with a particular emission factor, the level of uncertainty in the emission estimate increases. Thus it is important to determine not only what the value of the emission factor is, but also to review carefully the conditions associated with that emission factor. If the conditions in your process do not correspond to those identified with the emission factor, emission estimates using that factor will be very uncertain.

The most comprehensive and reliable source of emission factors for thousands of different processes is AP-42, a publication of the U.S. EPA.[1] The introduction to AP-42 indicates that the emission factors contained therein "are simply averages of all available data of acceptable quality, . . ." My experience is that emission factors in AP-42 tend to be conservative. That is, they tend to overestimate the emissions for a particular process. Perhaps an explanation for this experience of overestimation is that the data in AP-42 are usually several years old, and consequently do not include newer equipment which will usually have lower emissions. An example of this can be seen in Figure 2-2. For utility boilers, the emission factor given in AP-42 of 550 lb/mmscf of natural gas use is well above the New Source Performance Standard (NSPS) for electric utility boilers of 200 lb/mmscf. Thus a new boiler of this size could not receive permit approval if it emitted NO_x at a rate consistent with the AP-42 emission factor.

Table 2-1 Emission Factors for Uncontrolled Natural Gas Combustion[a,b]

Furnace size and type (10^6 Btu/hr heat input)	Particulate[c]		Sulfur dioxide[d]		Nitrogen oxides[e]		Carbon monoxide[f]		Total organic compounds	
	kg/10^6m^3	lb/10^6ft^3	kg/10^6m^3	lb/10^6ft^3	kg/10^6m^3	lb/10^6ft^3	kg/10^6m^3	lb/10^6ft^3	kg/10^6m^3	lb/10^6ft^3
Utility boilers (> 100)	16–80[g]	1–5[g]	9.6	0.6	8,800[h]	550[h]	640	40	28[i]	1.7[i]
Industrial boilers (10–100)	219	13.7	9.6	0.6	2,240	140	560	35	92[j]	5.8[j]
Commercial boilers (0.3–10)	192	12	9.6	0.6	1,600	100	320	21	128[k]	8.0[k]
Residential Furnaces (<0.3)	183	11.2	9.6	0.6	1,500	94	640	40	180[k]	11[k]

[a]Expressed as weight/volume fuel fired.
[b]Based on an average natural gas higher heating value of 8,270 kcal/m^3 (1000 Btu/ft^3).
[c]Includes both filterable and condensable PM. All particulate emissions can be assumed to be less than 10 microns (PM$_{10}$).
[d]Based on average sulfur content of natural gas, 4,600 g/10^6 Mm^3 (2,000 gr/10^6 scf)
[e]Expressed as NO$_2$. Tests indicate about 95 weight % NO$_x$ is NO$_2$.
[f]May increase 10 - 100 times with improper operation or maintenance.
[g]Only filterable PM included. No data are available on condensable PM.
[h]For tangentially fired units, use 4,400 kg/10^6 m^3 (275 lb/10^6 ft^3). At reduced loads, values will be lower.
[i]Methane comprises 17% of organic compounds.
[j]Methane comprises 52% of organic compounds.
[k]Methane comprises 34% of organic compounds.
Source AP-42, Section 1.4, Fifth Edition, 1/95.

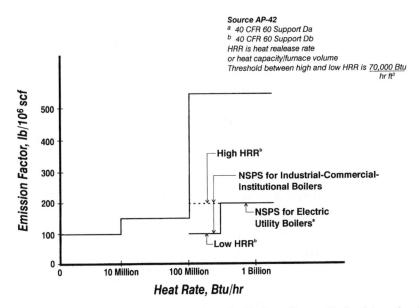

Figure 2-2 NO$_x$ emission factors for natural gas-fired boilers. (Source: Radian International, LLC.)

One useful feature of the emission factors presented in AP-42 is that they are graded. Factors that are considered to be especially reliable because of a large quantity of data points, usually more than 10, obtained using widely accepted test methods are given an A grade. Data from a single source test, or from sources that have highly variable emissions (or emission rates that are dependent upon several independent parameters), or utilize test methods that are not EPA approved, are rated E. As the quantity and quality of available data improves, higher grades are assigned. The grading reflects the professional judgment of the authors and reviewers of AP-42 sections concerning the reliability of the estimates that can be obtained using these factors.

Updates of AP-42 are made frequently. The most recent updates are available on a computer bulletin board that can be accessed from the Internet. The EPA home page is at http.//www.epa.gov. Access to AP-42 updates can be found from the home page by clicking on TTN.BBS.

Additional emission factors for toxic air pollutants can be found in the "Locating and Estimating . . ." series of reports produced by EPA.[2] Each report in the series focuses on a particular toxic air pollutant and identifies sources of emissions of that pollutant as well as emission factors from such sources based upon available measurements. Often the emission factors for a particular organic toxic air pollutant are presented as a fraction of the total hydrocarbon emissions for a process. For example, the emissions of formaldehyde from a combustion turbine operating on natural gas are given as 5% of the hydrocarbon emissions from the turbine.[3] From AP-42, the hydrocarbon emission factor for a natural gas-fired combustion turbine is 5 lb per million Btu heat input. Thus the formaldehyde emissions would be (Equation 2-3).

$$F_f = (0.05)\left[5\frac{\text{pounds}}{\text{million Btu heat input}}\right]$$

(2-3)

$$= 0.25\frac{\text{lb formaldehyde}}{\text{million Btu heat input}}$$

As with AP-42 estimates, careful attention must be paid to the conditions associated with a particular estimate to determine whether it will be applicable to calculating emissions for a particular source.

EMISSIONS FROM A BOILER USING AN EMISSION FACTOR

A distillate fired industrial boiler rated at 60 MMBtu/hr uses 2 million gallons per year of 0.25% sulfur distillate. What are its emissions of nitrogen oxides (NO_x) and formaldehyde?

Emission factors of criteria pollutants for distillate combustion in an industrial boiler are:[4]

Pollutant	Emission factor (lb/10^3 gallon)
Nitrogen oxides	20
Sulfur dioxide	142S*
PM_{10}	2
Carbon monoxide	5
Volatile organic compounds	0.2

The nitrogen oxide emission rate will be

$$\text{Emission rate} = 20\frac{\text{lb}}{10^3\text{gal}} \times 2000\frac{\text{thousand gallons}}{\text{year}}$$

$$= 40,000 \text{ lb/year} = 20 \text{ tons/year}$$

Emission factors for organic toxic air contaminant emissions from distillate fired in an industrial boiler include the following:[5]

Pollutant	Emission factor (lb/10^3 gallon)
Benzene	0.1863
Formaldehyde	1.7261
Cadmium	0.0015
Chromium (VI)	0.0002

The formaldehyde emission rate will be

$$\text{Emission rate} = 1.7261\frac{\text{lb}}{10^3\text{gal}} \times 2000\frac{\text{thousand gallons}}{\text{year}}$$

$$= 3452 \text{ lb/year} = 1.73 \text{ tons/year}$$

* S is % sulfur in fuel.

In some instances, emission estimates will be sought for a particularly unusual source. When a source is not included in AP-42 or other such documents, extrapolating from values that can be found should be considered highly uncertain, and in many instances may not even estimate the correct order of magnitude of emissions. In these instances, the sensitivity of permitting requirements to such uncertain estimates should be determined by conservatively assuming a tenfold increase in emissions over the values estimated. If such a high value renders the project unpermittable, source testing or other estimation techniques should be utilized in an effort to more accurately quantify emissions.

IV. ENGINEERING CALCULATIONS

Engineering calculations are used to estimate the emissions of a process using physical principles such as diffusion, heat and mass transfer, and fluid flow. Although estimates can be made from "first principles" without relying upon measurements, those estimates, commonly termed *engineering judgment* are usually not very accurate, as indicated in Figure 2-1. Another category of engineering calculations builds on the concept of emission factors discussed above. Where emissions are a function of several variables, use of physical principles enables the relationships among variables to be postulated, and then emission measurements can be used to determine the appropriate corrections to obtain accurate estimates.

STORAGE TANKS HAVE TWO CAUSES OF EMISSIONS

$$L_T = L_B + L_W$$

where L_T = total emissions, L_B = standing losses, and L_W = withdrawal losses

---

The most common emission unit category that uses engineering calculations is emission estimates for working and breathing losses from organic liquid storage tanks.[6] Working losses occur as a result of filling and emptying a tank. Breathing losses occur for standing liquids as a result of diurnal temperature variations. For this calculation, a correlation equation in which engineering principles have been incorporated into algebraic equations and numerical factors is used to estimate emissions. It is based upon the physical principles of heat transfer, mass balance, and liquid–vapor equilibrium applied to the storage tank. The application of these correlation equations is complicated but straightforward for single component liquids. However, for liquid mixtures, it is necessary to apply Raoult's Law to determine the composition of a multicomponent vapor above a liquid, and knowledge of chemical engineering is helpful.[7] The factors included in determination of tank losses include the size of the tank, the thruput, expressed as number of turnovers per year, the roof type (fixed, floating, etc.), the average vapor space, the color of the tank, the quality of the finish, and the geographic location.

The methods for calculating breathing and working losses have been incorporated into an EPA software package called TANKS PROGRAM 2.0, which is available on the EPA CHIEF bulletin board.[8] An illustration of the printout from the TANKS software is shown in Figure 2-3. Note that the terminology used in TANKS refers to working losses as "withdrawal losses" and breathing losses as "standing losses."

TANKS PROGRAM 2.0   07/13/95

EMISSIONS REPORT — DETAIL FORMAT   PAGE 1

TANK IDENTIFICATION AND PHYSICAL CHARACTERISTICS

```
Identification
    Identification No.:     Diesel
    City:                   Barrow, Alaska
    State:                  AK
    Company:
    Type of Tank:           Vertical Fixed Roof

Tank Dimensions
    Shell Height (ft):                30
    Diameter (ft):                    36
    Liquid Height (ft):               18
    Avg. Liquid Height (ft):           9
    Volume (gallons):             137071
    Turnovers:                        17
    Net Throughput (gal/yr):     2266666

Paint Characteristics
    Shell Color/Shade:      Red/Primer
    Shell Condition:        Good
    Roof Color/Shade:       Red/Primer
    Roof Condition:         Good

Roof Characteristics
    Type:                   Cone
    Height (ft):                    0.00
    Radius (ft) (Dome Roof):        0.00
    Slope (ft/ft) (Cone Roof):    0.0625

Breather Vent Settings
    Vacuum Setting (psig):         -0.01
    Pressure Setting (psig):        0.01
```

Meteorological Data Used in Emission Calculations: Barrow, Alaska

Figure 2-3 Fixed roof tank emissions using TANKS 2.0 Software. (Source: EPA Office of Air Quality Planning and Standards, Technology Transfer Network (TTN), (919) 541-5742.)

Another application of engineering calculations is to determine fugitive emissions from paved and unpaved roads and from piles. Several variables are important in calculating these particulate emissions. Detailed descriptions of the techniques used are found in Section 11 of AP-42.

```
                    TANKS PROGRAM 2.0                07/13/95
             EMISSIONS REPORT - DETAIL FORMAT        PAGE 2
               LIQUID CONTENTS OF STORAGE TANK


                               Liquid
                 Daily Liquid Surf.    Bulk
                 Temperatures (deg F)  Temp.   Vapor Pressures (psia)   Vapor
                                      (deg F)                           Mol.
Mixture/Component  Month  Avg.  Min.  Max.     Avg.    Min.    Max.     Weight
-------------------------------------------------------------------------------

Diesel             All    15.61 9.98  21.24    13.34   0.0130  0.0010   0.0250  130.000



            Liquid  Vapor
            Mass    Mass    Mol.    Basis for Vapor Pressure
            Fract.  Fract.  Weight  Calculations
            -----------------------------------------------------
```

Figure 2-3 *(Continued).*

TANKS PROGRAM 2.0   07/13/95

EMISSIONS REPORT — DETAIL FORMAT   **PAGE 3**

DETAIL CALCULATIONS (AP-42)

Annual Emission Calculations

Standing Losses (lb):	123.6349
Vapor Space Volume (cu ft):	21757.10
Vapor Density (lb/cu ft):	0.0003
Vapor Space Expansion Factor:	0.047676
Vented Vapor Saturation Factor:	0.985486

Tank Vapor Space Volume	
Vapor Space Volume (cu ft):	21757.10
Tank Diameter (ft):	36
Vapor Space Outage (ft):	21.38
Tank Shell Height (ft):	30
Average Liquid Height (ft):	9
Roof Outage (ft):	0.38

Roof Outage (Cone Roof)	
Roof Outage (ft):	0.38
Roof Height (ft):	0.000
Roof Slope (ft/ft):	0.06250
Shell Radius (ft):	18

Vapor Density	
Vapor Density (lb/cu ft):	0.0003
Vapor Molecular Weight (lb/lb-mole):	130.000000
Vapor Pressure at Daily Average Liquid	
Surface Temperature (psia):	0.013000
Daily Avg. Liquid Surface Temp.(deg. R):	475.28
Daily Average Ambient Temp. (deg. R):	468.67
Ideal Gas Constant R	
(psia cuft /(lb-mole-deg R)):	10.731
Liquid Bulk Temperature (deg. R):	473.01
Tank Paint Solar Absorptance (Shell):	0.89
Tank Paint Solar Absorptance (Roof):	0.89
Daily Total Solar Insolation	
Factor (Btu/sqftday):	595.00

Vapor Space Expansion Factor	
Vapor Space Expansion Factor:	0.047676
Daily Vapor Temperature Range (deg.R):	22.53
Daily Vapor Pressure Range (psia):	0.024000
Breather Vent Press. Setting Range(psia):	0.02
Vapor Pressure at Daily Average Liquid	
Surface Temperature (psia):	0.013000
Vapor Pressure at Daily Minimum Liquid	
Surface Temperature (psia):	0.001000
Vapor Pressure at Daily Maximum Liquid	
Surface Temperature (psia):	0.025000
Daily Avg. Liquid Surface Temp. (deg R):	475.28
Daily Min. Liquid Surface Temp. (deg R):	469.65
Daily Max. Liquid Surface Temp. (deg R):	480.91
Daily Ambient Temp. Range (deg.R):	10.70

Figure 2-3 *(Continued).*

TANKS PROGRAM 2.0   07/13/95
EMISSIONS REPORT — DETAIL FORMAT   PAGE 4
DETAIL CALCULATIONS (AP-42)

```
Annual Emission Calculations
Vented Vapor Saturation Factor
   Vented Vapor Saturation Factor:            0.985486
   Vapor Pressure at Daily Average Liquid
   Surface Temperature (psia):                0.013000
   Vapor Space Outage (ft):                     21.38

Withdrawal Losses (lb):                         91.2063
   Vapor Molecular Weight (lb/lb-mole):      130.000000
   Vapor Pressure at Daily Average Liquid
   Surface Temperature (psia):                0.013000
   Annual Net Throughput (gal/yr):            2266666
   Turnover Factor:                            1.0000
   Maximum Liquid Volume (cuft):                18322
   Maximum Liquid Height (ft):                     18
   Tank Diameter (ft):                             36
   Working Loss Product Factor:                  1.00

Total Losses (lb):                             214.84
```

Annual Emissions Report

| Liquid Contents | Losses (lbs.): | | |
	Standing	Withdrawal	Total
Diesel	123.63	91.21	214.84
Total:	123.63	91.21	214.84

Figure 2-3 *(Continued)*.

V. MATERIAL BALANCE

A material balance utilizes the conservation of mass to estimate emissions. Using a material balance assumes that no mass of the pollutant being estimated is created or destroyed. Hence this technique is most suited for noncombustion processes such as degreasers or paint spray booths where volatile organic emissions are the result of evaporation of volatile organic compounds (VOC) entering the system. The material balance is an accounting of the materials entering and leaving a defined system (Equation 2-4):

$$\text{Emissions} = \text{Material in} - \text{Material out} \qquad (2\text{-}4)$$
$$\text{(in non-air streams)}$$

In some instances, material balances are made for an entire facility, such as to estimate the emissions of a toxic pollutant to complete Form R reporting under the requirements of the Emergency Planning and Community Right-to-Know Act (EPCRA), Section 313. However, facility-wide material balance estimates have many opportunities for inaccuracies because it is difficult to identify all of the sources and releases or disposal of the pollutant.

The most accurate use of material balances is for simple processes such as surface coating or solvent use, or for emission estimates of metals that enter a system as part of a fuel or feedstock. A material balance used to estimate VOC and chromium emissions from a paint booth is shown below. Because the emissions are the difference between material entering the paint spray booth as components of the paint and the material leaving the booth on the painted product or captured in pollution control equipment, the estimates will tend to be conservative. That is, the emission estimate will tend to be above the actual emissions from the process. This is because losses to the walls of the booth or in not-quite-empty paint cans are usually not accounted for, and therefore allocated to air emissions.

The most common assumption for emissions of VOCs from painting operations is that 100% of VOCs in the paint are released as air emissions. Although these emissions will not all be from the point of the spray booth, VOCs in the drying paint as well as that left in the residue in the can will eventually be emitted to the air. The only opportunities to reduce these VOC emissions are through reductions in the content of VOCs in the paint or through VOC control devices such as carbon adsorption or incineration of the VOCs in the exhaust airflow from the booth.[9]

MATERIAL BALANCE FOR A SPRAY BOOTH

A surface coating spray booth has a coating usage of 500 gal/year. The coating used has a density of 8 lb/gal and a concentration of 0.08% hexavalent chromium. The transfer efficiency of the spray gun used is 40%. The control efficiency of mat filters on the booth is 80%.

What is the hexavalent chromium emission rate from the booth?

$$\text{Emission rate} = \text{Mass in} - \text{Mass removed}$$

Mass of hexavalent chromium [Cr(VI)] into the spray booth

$$= 500 \frac{\text{gallons}}{\text{year}} \times 8 \frac{\text{pounds}}{\text{gallon}} \times 0.0008 \frac{\text{pounds Cr(VI)}}{\text{pounds coating}}$$

$$= 3.2 \frac{\text{pounds}}{\text{year}}$$

Mass of Cr(VI) removed from the spray booth includes the amount transferred to the surface being painted and the amount of overspray captured by the mat filters.

Mass of Cr(VI) transferred to the surface being painted

$$\text{Transferred} = 3.2 \frac{\text{pounds}}{\text{year}} \times 0.4$$

$$\text{Transferred} = 1.28 \frac{\text{pounds}}{\text{year}}$$

Mass of Cr(VI) oversprayed is the amount entering the booth minus the amount transferred to the surface being painted

$$\text{Overspray} = 3.2 \frac{\text{pounds}}{\text{year}} - 1.28 \frac{\text{pounds}}{\text{year}}$$

$$\text{Overspray} = 1.92 \frac{\text{pounds}}{\text{year}}$$

Mass of Cr(VI) captured in mat filters is 80% of the overspray

$$\text{Captured} = 1.92 \frac{\text{pounds}}{\text{year}} \times 0.8$$

$$\text{Captured} = 1.54 \frac{\text{pounds}}{\text{year}}$$

The emissions will be the amount of Cr(VI) entering the booth minus the amount transferred to the surface being coated minus the amount captured in the mat filters

$$\text{Emissions} = 3.2 \frac{\text{pounds}}{\text{year}} - 1.28 \frac{\text{pounds}}{\text{year}} - 1.54 \frac{\text{pounds}}{\text{year}}$$

$$\text{Emissions} = 0.38 \frac{\text{pounds}}{\text{year}}$$

VI. SOURCE TESTING

The most accurate method for estimating emissions from a process or emission unit is source testing. Source testing measures the concentration of a pollutant in the flow from a stack or vent, measures the flow rate, and obtains the emission estimate as the product of these two measurements (Equation 2-5):

$$\text{Emissions} = (\text{Pollutant concentration}) * (\text{Flow rate}) \qquad (2\text{-}5)$$

However, several things can affect the accuracy of this technique. They include the following:

- Use of approved source test and analysis methods
- Replicate process operations in the test conditions
- Maintenance of steady state operation
- Measurements well above detection limits without interference

A. Use Approved Source Test and Analysis Methods

First, approved source test and analysis methods must be used. EPA has developed source test protocols that are accepted by most state and local jurisdictions and have been used long enough that their application has little uncertainty. The California Air Resources Board (CARB) has also developed test protocols that may differ slightly from EPA methods. The CARB protocols should be used in California for the results to be accepted by local regulatory agencies. Analysis methods are used in conjunction with source test methods, and frequently the first step in the analysis is capturing the pollutant in an appropriate medium to assure an accurate measurement. Analysis methods are described in SW-846,[10] a four-volume set of methods that have been approved by EPA as well as in Appendix M to 40 CFR Part 60.6.

One important issue in source testing is to be sure that isokinetic sampling is used when called for in a protocol. Isokinetic sampling requires that the velocity of flow into a sampling probe is the same as the velocity in the free stream being sampled. It is usually required when testing is for particulate matter or organic contaminants that could be present as particulate (semi-volatiles). The reason for this is illustrated in Figure 2-4. For a stream with nonuniform particle sizes, when the velocity into the probe is greater than the velocity in the free stream, streamlines are pulled into the probe and more gas per unit area enters the probe than is present in the stack. However, because they have a large mass compared to gas molecules, large particulates in the free stream cannot turn as readily as the gas molecules, and consequently fewer small particulates are collected than large particles, thus biasing the test data upward.

If there is a lower velocity in the probe than in the free stream, streamlines curve out, and less gas enters the probe per unit area. However, the large particulates cannot turn as easily, and this results in a higher large particulate concentration measurement than exists in the free stream.

Accurate particulate concentrations are measured only when the velocity of gas into the probe is the same as in the free stream. This condition is called isokinetic sampling.

Source tests and analysis methods do continue to change, and for some test conditions or pollutants, there may not be an approved method. For example, the ambient standard for particulate matter less than 10 μm (PM_{10}) was established in 1987. At the present time, there is no approved method for measurement of PM_{10} in a flow containing water vapor. No technique has been demonstrated that will distinguish between small particles and water vapor.

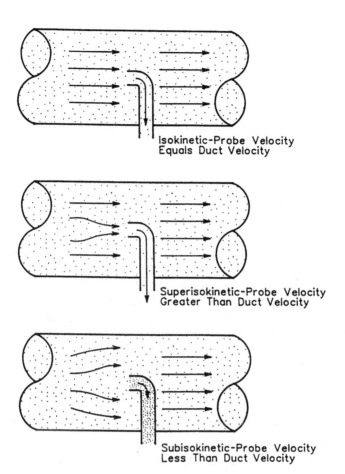

Isokinetic-Probe Velocity
Equals Duct Velocity

Superisokinetic-Probe Velocity
Greater Than Duct Velocity

Subisokinetic-Probe Velocity
Less Than Duct Velocity

Figure 2-4 Consequences of non-isokinetic flow during source testing. (Source: Radian International, LLC.)

There may also not be approved methods for some hazardous air pollutants. Other test methods have been developed, such as those developed by the Society for Industrial Hygiene. These methods are not recommended for source testing, but are used when approved methods are not available.

B. Replicate Process Operations in the Test Conditions

For a source test to be used as the basis for a process emission estimate, the conditions of operation of the process must have been replicated in the source test. Although this point may seem obvious, it is a common difficulty associated with source testing. In some jurisdictions, an agency will not approve the use of source test results unless agency staff were present for the test to be sure that the process was in normal operation during source testing.

Another problem that can occur during source testing is that process parameters are not recorded. It is not necessary to keep a record of process parameters, but if there is a need or benefit to developing emission factors based on a source test, a numerical record of process parameters must be available.

In some instances, it is not possible to duplicate operating conditions in a source test. This might be the case for batch operations in which the tests duration prescribed in a test protocol cannot be achieved during a batch. When this is the case, a shorter test duration may be used, with the potential for significantly increased detection limits, or testing can be conducted for several batches. In this latter case, care must be taken to keep samples from deteriorating or being contaminated between batches.

C. Maintenance of Steady-State Operation

The flow during a source test must be consistent with the intended purpose of the test. If steady state emissions are the goal of the testing, the flow must be uniform and steady during the test. If the flow rate or uniformity varies, a more involved testing protocol will be necessary, increasing the expense of the source test. It is particularly difficult to conduct source tests when there are spikes or slugs of concentration. A sample that includes unperceived slugs may badly overestimate emissions, and one that avoids them completely may similarly underestimate emissions. Usually process engineers have a good understanding of whether there are spikes in the operation. For these sources, the process engineer and testing engineer must have a clear and mutual understanding of the data uses to design an appropriate test.

D. Measurements Well Above Detection Limits Without Interference

Detection limits are often the most troublesome consideration in source testing. Failure to consider the detection limit needed for a source test can completely invalidate the test or provide results that prohibit approval of the process permit. One example of this is in testing for hazardous air pollutants. Hexavalent chromium is a highly carcinogenic pollutant, and it is suspected of being present in any gas stream that contains chromium. However, the test methods available for measuring hexavalent chromium have the same detection limits as those for measuring total chromium. Also, hexavalent chromium is present in low quantities in the sampling reagent. As a result, a source test for hexavalent chromium may not be sensitive enough to demonstrate that less than all of the chromium present is hexavalent. In instances where chromium concentrations are small, use of engineering calculations or data from similar tests on other equipment at higher concentrations may be more useful than a hexavalent chromium source test.

VII. OTHER EMISSION ESTIMATION TOPICS

Several other topics can be important in making emission estimates. These include use of F factors to estimate heat input to a combustion process, considerations of startup and shutdown in estimating emissions, estimating fugitive emissions from process piping systems, and making corrections to emission estimates to obtain results in different units of dimension.

A. Use of F Factors

As discussed in the previous section, it is important to collect process data in order to use source test results to obtain an emission factor. However, there may be occasions when process data are not obtained, and EPA has developed a method for determining heat input to a combustion process if the flue gas flow rate is known. The ratio of flue gas volume to heat input is known as an F factor, and it is determined from combustion calculations.

To obtain emission factors using F factors, the type of fuel used, the amount of oxygen or carbon dioxide in the flue gas, the moisture in the flue gas, and the pollutant concentration must be known. For the discussion below, a dry flue gas is assumed. To correct for moisture content, see subsection D below. A more detailed discussion with information on application of F factors for wet gas streams and for fuels other than those shown in Table 2-2 is taken from 40 CFR Part 60, Appendix A, Method 19.

Table 2-2 F-Factors for Determining Emission Factors from Measured Pollutant Concentrations[a]

Fuel	F_d (measuring oxygen)		F_c (measuring carbon dioxide)	
	dscm/J	dscf/10⁶ Btu	scm/J	scf/10⁶ Btu
Coal				
Anthracite[b]	2.71×10^{-7}	10,100	0.530×10^{-7}	1,970
Bituminous[b]	2.63×10^{-7}	9,780	0.484×10^{-7}	1,800
Lignite	2.65×10^{-7}	9,860	0.513×10^{-7}	1,910
Oil[c]	2.47×10^{-7}	9,190	0.383×10^{-7}	1,420
Gas:				
Natural	2.34×10^{-7}	8,710	0.287×10^{-7}	1,040
Propane	2.34×10^{-7}	8,710	0.321×10^{-7}	1,190
Butane	2.34×10^{-7}	8,710	0.337×10^{-7}	1,250
Wood	2.48×10^{-7}	9,240	0.492×10^{-7}	1,830
Wood bark	2.58×10^{-7}	9,600	0.516×10^{-7}	1,920
Municipal solid waste	2.57×10^{-7}	9,570	0.488×10^{-7}	1,820

[a]Determined at standard conditions: 20°C (68°F) and 750 mm Hg (29.92 in. Hg).
[b]As classified according to ASTM D388-77.
[c]Crude, residual, or distillate.
Source: 40 CFR Part 60, Appendix A, Method 19

The equation to obtain an emission factor, E, when measurements are on a dry basis for both oxygen and the pollutant is

$$E = C_d F_d \frac{20.9}{20.9 - \%O_{2d}} \qquad (2\text{-}7)$$

where C_d = pollutant concentration, F_d = F factor for the fuel being used (measuring oxygen), and O_{2d} = percent oxygen in the flue gas on a dry basis.

When measurement has been taken of CO_2 concentration in the flue gas instead of oxygen, the emission factor, E, can be found from

$$E = C_d F_c \left(\frac{100}{\%CO_{2d}} \right) \tag{2-8}$$

where C_d = pollutant concentration, F_c = F factor for the fuel being used (measuring carbon dioxide), and CO_{2d} = percent CO_2 in the flue gas on a dry basis.

EMISSION FACTOR FOR MERCURY USING AN F FACTOR

A concentration of 2×10^{-9} pounds per cubic foot of mercury is measured in the flue gas from distillate combustion. If the oxygen level in the stream is 3%, calculate the resulting emission factor for mercury.

$$E = C_d F_d \left[\frac{20.9}{(20.9 - \%O_{2d})} \right]$$

$$E = 2 \times 10^{-9} \frac{\text{pounds}}{\text{ft}^3} \times 9190 \frac{\text{ft}^3}{10^6 \text{Btu}} \left[\frac{20.9}{(20.9 - 3)} \right]$$

$$E = 2 \times 10^{-9} \frac{\text{pounds}}{\text{ft}^3} \times 9190 \frac{\text{ft}^3}{10^6 \text{Btu}} \left[\frac{20.9}{17.9} \right]$$

$$E = 2 \times 10^{-9} \frac{\text{pounds}}{\text{ft}^3} \times 9190 \frac{\text{ft}^3}{10^6 \text{Btu}} \times 1.17$$

$$E = 21 \frac{\text{lb}}{10^{12} \text{Btu}}$$

B. Startups, Shutdowns, and Upsets

Any piece of equipment, process, or facility experiences startups and shutdowns, and over its lifetime, will also operate in abnormal or upset modes. Often the emissions from the source are different during startups, shutdowns, and upsets. One common reason for different emissions during startup is because the equipment or process must warm up to its operating temperature. During this warmup period, a combustor usually has a lower temperature and combustion efficiency so that CO and hydrocarbon emissions are higher and NO_x emissions are lower than during steady state operation. If combustion efficiency is extremely low, soot may form and create significantly increased particulate emissions compared to steady state operation.

For some processes, pollution control equipment may need to warm up, and control efficiencies may be near zero during the early part of the warmup. This is the situation for the Selexol process, which, during normal operation provides 98 to 99% control of acid gases such as H_2S and CO. During startup, the process provides much lower control, which can result in very high emissions of sulfur and carbon dioxide.

Several different events can cause upsets in a process or its pollution control equipment. The failure of a bag in a baghouse or arcing in an electrostatic precipitator can result in significant increases in particulate emissions. A pressure excursion in a process vessel can result in a relief valve opening and the release of process chemicals as air pollutants.

Historically, air permits have excluded startup, shutdown, and upset emissions from allowable emission rates. However, this policy is changing. Concern for short-term health effects from upsets followed the 1987 Union Carbide incident in Bhopal, India in which a plant upset resulted in the release of methyl isocyanite that killed or injured thousands of people. Consequently, both federal and state statutes have required estimation of upset emissions and assessments of the potential for damage to public health from those emissions. This is discussed further in Chapter 6.

In some jurisdictions, regulatory agencies may require estimation of short-duration emissions associated with startups and shutdowns and modeling of those emissions to verify that short-term standards are not exceeded. This is more likely to be the case for facilities or processes in which startups, shutdown, or upsets are frequent. Several years ago in New Mexico, regulations were promulgated for upsets after it was found that several large power plants were operating in an "upset mode" for a significant percentage of total operating time, and emitting many times allowable steady-state levels of particulates.

Estimating emissions during startup, shutdown, or upset periods utilizes the same emission estimation methods described above. However, data are not as plentiful for these operating conditions, and consequently emission estimates are usually not as accurate. It may be necessary to rely upon engineering judgment more frequently for these estimates.

For sulfur dioxide emissions, mass balance can be quite useful for estimating emissions during startup or shutdown when sulfur dioxide pollution control equipment does not function at design efficiency. In those instances, all or a portion of the sulfur entering the process can be assumed to be released as sulfur dioxide.

Since most jurisdictions have strict regulations against visible emissions, regardless of the cause of the emissions, many equipment vendors will guarantee that no soot will be formed during startup or shutdown, and particulate emissions can be assumed to be the same as for steady-state operation.

Nitrogen dioxide emissions are dependent upon operating temperature and residence time at the peak combustion temperature for fuels containing little or no nitrogen. Since warmup means that the combustor is not yet up to operating temperature, NO_x emissions will be lower than steady-state levels.

Carbon monoxide levels may be much higher during startup because of low combustion efficiency. Steady-state emissions can be assumed to be the lower limit of CO startup emissions unless supporting data are available to indicate otherwise. Vendor data and source tests may be the only basis available for CO emission estimates on startups.

C. Fugitive Emission Estimates

This subsection discusses fugitive emission estimates for piping systems typical of chemical processing plants or oil refineries. Fugitive emissions are important from these facilities because although the emissions from a single pipe junction or valve are very small, the large number of potentially leaking sources multiplies these small rates to substantial total emissions. Fugitive emissions are also very important in hazardous waste incineration facilities in which the liquids or gases being piped are often toxic, and extremely small emission rates can result in significant public health impacts.

Fugitive emissions are estimated by first counting the number of valves, pump seals, compressor seals, pressure relief valves, flanges, sampling connections, and open-ended lines in a process or facility. Open-ended lines are used to tap into a process and are closed by a valve. Emissions from open-ended lines are thus the emissions due to leaks through a closed valve.

Several methods have been developed to estimate equipment leaks — fugitive emissions — from chemical processing units. Two that are discussed here use:

- Average emission factors
- Leak/no-leak emission factors

Average emission factors are used where no measurements are available and no inspection and maintenance procedures are proposed. These are shown in Table 2-3. If organic vapor analyzers (OVA) have been used to determine the number of devices leaking, the leak/no-leak factors can be used. These are shown in Table 2-4. These factors can also be used for future construction by estimating the number of leakers based upon the frequency of inspection and maintenance proposed. Quarterly inspections are usually assumed to result in a 10% leak rate. More sophisticated emission factors have been developed for operating equipment with more detailed leak measurements available. These more detailed leak measurements use equipment that measures the concentration of organic vapors more accurately than an OVA. More accurate, and hence less conservative, fugitive emission estimates can be made from use of these more sophisticated correlations.[11] All of these emission factors are based on measurements that have been made of emissions from each of these devices. The calculation of fugitive emissions is illustrated below.

Table 2-3 Average Emission Factors for Fugitive Emissions

Equipment	Service	Emission factor (kg/hr/source)
Valves	Gas	0.0056
	Light liquid	0.0071
	Heavy liquid	0.00023
Pump seals	Light liquid	0.0494
	Heavy liquid	0.0214
Compressor seals	Gas/vapor	0.228
Pressure relief seals	Gas/vapor	0.104
Flanges	All	0.00083
Open-ended lines	All	0.0017
Sampling connections	All	0.0150

Source: U.S. EPA, Protocols for Generating Unit-Specific Emission Estimates for Equipment Leaks of VOC and VHAP. EPA-450/3-88-010, U.S. Environmental Protection Agency, Office of Air Quality Planning and Standards, Research Triangle Park, N.C., October, 1988

FUGITIVE EMISSIONS FROM A CATALYTIC CRACKER

A catalytic cracker in an oil refinery has the component count shown in the table below. Determine the fugitive emissions from this process using both average emission factors and leaking/nonleaking emission factors.

	Gas/vapor	Light liquid	Heavy liquid
Pump seals		47/3*	3/1
Valves	625/19	1180/3	64/0
Pressure relief valves	31/1		
Open-ended lines	54/3	192/5	32/1
Compressor seals	4/0		
Sampling connections	25/0	35/0	10/0
Flanges	830/6	1620/11	430/3

* Total/number found leaking

Average Emission Factors

Based on the emission factors in Table 2-3, the emissions for pump seals in light liquid service would be

$$E(PS)_{LL} = (\text{No. of pump seals}) \times (\text{EF for pump seals})$$

$$E(PS)_{LL} = 47 \text{ sources} \times 0.0494 \frac{\text{kg}}{\text{hr source}}$$

$$E(PS)_{LL} = 2.32 \frac{\text{kg}}{\text{hr}}$$

If the equipment operated year around, the annual emissions would be

$$E(PS)_{LL} = 2.32 \frac{kg}{hr} \times 8760 \frac{hr}{year}$$

$$E(PS)_{LL} = 2.03 \times 10^4 \frac{kg}{year} \times \frac{1 \text{ ton}}{907.2 \text{ kg}}$$

$$E(PS)_{LL} = 22.4 \frac{tons}{year} \text{ from pump seals in light liquid service.}$$

Total pump seal emissions would be the sum of the emissions from seals in light liquid service and heavy liquid service.

$$E(PS) = E(PS)_{LL} + E(PS)_{HL}$$

$$E(PS) = 22.4 \frac{tons}{year} + 3 \text{ sources} \times 0.0214 \frac{kg}{hr \text{ source}} \times 8760 \frac{hr}{year} \times \frac{1 \text{ ton}}{907.2 \text{ kg}}$$

$$E(PS) = 22.4 \frac{tons}{year} + 0.6 \frac{tons}{year}$$

$$E(PS) = 23.0 \frac{tons}{year}$$

Leaking/Nonleaking Emission Factors

Based on the emission factors in Table 2-4, the emissions from pump seals in light liquid service would be

$$E(PS)_{LL} = (\text{no. of nonleaking pump seals}) \times (\text{nonleaking EF}) +$$

$$(\text{no. of leaking pump seals}) \times (\text{leaking EF})$$

$$E(PS)_{LL} = 44 \text{ sources} \times 0.0120 \frac{kg}{hr \text{ source}} \times + 3 \text{ sources} \times 0.437 \frac{kg}{hr \text{ source}}$$

$$E(PS)_{LL} = 0.53 \frac{kg}{hr} + 1.31 \frac{kg}{hr}$$

$$E(PS)_{LL} = 1.84 \frac{kg}{hr}$$

The annual emissions from pump seals for light liquid service would be

$$E(PS)_{LL} = 1.84 \frac{kg}{hr} \times 8760 \frac{hr}{year} \times \frac{1 \text{ ton}}{907.2 \text{ kg}}$$

$$E(PS)_{LL} = 17.8 \frac{tons}{year}$$

The annual emissions from pump seals for heavy liquid service would be

$$E(PS)_{HL} = 2 \text{ sources} \times 0.0135 \frac{kg}{hr \text{ source}} + 1 \text{ sources} \times 0.3885 \frac{kg}{hr \text{ source}}$$

$$E(PS)_{HL} = 0.27 \frac{kg}{hr} + 0.3885 \frac{kg}{hr}$$

$$E(PS)_{HL} = 0.42 \frac{kg}{hr} \times 8760 \frac{hr}{year} \times \frac{1 \text{ ton}}{907.2 \text{ kg}}$$

$$E(PS)_{HL} = 4.0 \frac{tons}{year}$$

Total emissions using the leaking/nonleaking emission factors would be

$$E(PS) = E(PS)_{LL} + E(PS)_{HL}$$

$$E(PS) = 17.8 \frac{tons}{year} + 4 \frac{tons}{year}$$

$$E(PS) = 21.8 \frac{tons}{year}$$

The emission estimate using the leaking/nonleaking emission factors is about 5 percent less than the average emission factor estimate. The difference will vary with the number of leaking components.

D. Emission Estimate Corrections

Often, after emission calculations have been completed, it is necessary to compare emission rates or concentrations to some standard or permit condition. When doing so, a common basis for comparison must be used. Often permit conditions are referenced to standard conditions that are defined as part of agency rules and regulations. An interesting feature of such definitions is that they are not consistent among agencies. Standard pressure is always 1 atmosphere (760 mm Hg), but standard temperature varies, with the most common choices being 60 and 70°F, although 20°C (68°F) and 0°C have also been used. The EPA has used standard conditions as 60°F and 20°C (68°F).[13] In this text, we will use 20°C (68°F) for standard temperature.

Table 2-4   Leaking and Nonleaking Emission Factors for Fugitive
            Emissions (kg/hr/source)

Equipment	Service	Leaking (≥10,000 ppm) emission factor	Non-leaking (≥10,000 ppm) emission factor
Valves	Gas[a]	0.0451	0.00048
	LL[b]	0.0852	0.00171
	HL[c]	0.00023[d]	0.00023
Pump seals	LL	0.437	0.0120
	HL	0.3885	0.0135
Compressor seals[e]	Gas	1.608	0.0894
Pressure relief seals	Gas	1.691	0.0447
Flanges	All	0.0375	0.0006
Open-ended lines	All	0.01195	0.00150

[a]The leaking and non-leaking emission factors for valves in gas/vapor
 service are based upon the emission factors determined for gas valves
 in ethylene, cumene, and vinyl acetate units during the SOCMI
 Maintenance Study.
[b]LL = light liquid service. Vapor pressure greater than 0.1 psi at 100°F.
[c]HL - heavy liquid service. Usually kerosene is considered the lightest of
 the heavy liquids.
[d]Leaking emission factor assumed equal to non-leaking emission factor
 since the computed leaking emission factor (0.0005 kg/hr/source) was
 less than non-leaking emission factor.
[e]Emission factor reflects existing control level of 60% found in the industry;
 control is through the use of barrier fluid/degassing reservoir/vent-to-flare
 or other seal leakage capture system.
Source: U.S. EPA, Protocols for Generating Unit-Specific Emission
Estimates for Equipment Leaks of VOC and VHAP. EPA-450/3-88-010, U.S.
Environmental Protection Agency, Office of Air Quality Planning and
Standards, Research Triangle Park, N.C., October, 1988

1. Dry Gas Flow Calculation

Flow rates or emission rates often need to be presented in units of dry
standard cubic feet per minute. This requires not only correcting to standard
conditions, but also correcting for the moisture in the flow. The justification
for calculating flow on a dry basis is that water is not considered a pollutant,
and including the quantity of water in the flow will produce a lower concentration of pollutant because of the dilution. Therefore the dry basis is useful as
a means of comparison.

To determine dry basis, the humidity and gas temperature must be known.
Often humidity is presented as wet bulb temperature, and with the wet bulb and
gas temperature (known for moisture calculations as dry bulk temperature), a
psychometric chart can be used to obtain percent moisture. With percent moisture
information, dry gas flow can be determined from Equation 2-8.

$$\text{Dry gas flow} = \left(\frac{100 \pm \% \text{ moisture}}{100}\right) (\text{Wet gas flow}) \qquad (2\text{-}8)$$

2. Flow at Standard Temperature and Pressure (STP)

Flow at standard temperature and pressure (STP) can be determined provided actual temperature and pressure of the flow are known. Using EPA's standard conditions 20°C (68°F), flow at STP becomes (Equation 2-9):

$$\text{Flow @ STP} = (\text{Dry gas flow})\left(\frac{\text{Pm}}{14.7 \text{ psi}}\right)\left(\frac{528°\text{R}}{\text{Tm}}\right) \qquad (2\text{-}9)$$

where Pm is the measured pressure in psi and Tm the measured temperature in °R (°R = 460 + °F).

3. Conversion from ppm to $\mu g/m^3$

Conversion from ppm to $\mu g/m^3$ requires that the molecular weight of the gas is known. The equation is, for standard temperature = 20°C (68°F), (Equation 2-10)

$$\text{Conc}\left[\frac{\mu g}{m^3}\right] = 41.6 \times \text{MW} \times \text{Conc[ppm]} \qquad (2\text{-}10)$$

CONVERSION FROM *PPMV* TO $\mu g/m^3$ REQUIRES THAT YOU KNOW THE MOLECULAR WEIGHT OF THE POLLUTANT

For carbon monoxide the ambient air quality standard is 9.0 ppm (8-hr average). What is the concentration in $\mu g/m^3$?

$$\text{Conc}\left[\frac{\mu g}{m^3}\right] = 41.6 \times \text{MW} \times \text{Conc[ppm]}$$

The molecular weight of carbon monoxide is 28, hence

$$\text{Conc}\left[\frac{\mu g}{m^3}\right] = 41.6 \times 28 \times \text{Conc[ppm]}$$

$$\text{Conc}\left[\frac{\mu g}{m^3}\right] = 1164.8 \times 9.0$$

$$\text{Conc}\left[\frac{\mu g}{m^3}\right] = 10,483 \frac{\mu g}{m^3}$$

Since the primary standard is only given at two significant figures, the conversion needs to be rounded to two significant figures.

$$\text{Conc}\left[\frac{\mu g}{m^3}\right] = 10,000\,\frac{\mu g}{m^3}$$

$$\text{Conc} = 10\,\frac{mg}{m^3}$$

4. Corrections for Percent O₂

Corrections for percent O_2 are often needed for referencing combustion source emissions. For boilers, comparisons are made at 3% oxygen, consistent with the near stoichiometric conditions in boiler combustors. For combustion turbines, comparisons are made at 15% oxygen, consistent with the high excess air flow in a combustion turbine. To correct for concentrations at other oxygen levels, the following equation is used (Equation 2-12):

$$\text{conc}[3\%\ O_2] = \text{conc}[m\%\ O_2]\left[\frac{20.9 - 3.0}{20.9 - m\%\ O_2}\right] \qquad (2\text{-}12)$$

where m% O_2 is the nonstandard oxygen concentration. To calculate the concentration at 15% O_2, substitute 15.0 for 3.0 in Equation 2-12.

BOILER FLUE GAS CONCENTRATIONS ARE USUALLY CORRECTED TO 3% OXYGEN

The particulates in a stack gas are measured as 0.045 grains/dry standard cubic foot (dscf) with a stack gas oxygen concentration of 7% by volume. What is the particulate concentration at 3% oxygen?

$$\text{Conc}_{3\%} = \text{Conc}_{X\%}\left[\frac{20.7\% - 3\%}{20.7\% - X\%}\right]$$

$$= 0.045\,\frac{\text{grains}}{\text{dscf}}\left[\frac{20.7\% - 3\%}{20.7\% - 7\%}\right]$$

$$= 0.045\,\frac{\text{grains}}{\text{dscf}}\left[\frac{17.7\%}{13.7\%}\right]$$

$$\text{Conc}_{3\%} = 0.058\,\frac{\text{grains}}{\text{dscf}}\ @\,3\%\ O_2$$

VIII. PROBLEMS

1. An industrial boiler is rated at 60 million Btu/hr and can operate on either natural gas or diesel fuel. The only time the boiler runs on diesel is for 10 hours per month to test the diesel's capability. The boiler operates for 50 weeks per year including the time testing diesel. Assume natural gas has 1000 Btu/ft³ and diesel has 130,000 Btu/gal. Diesel fuel has the same emission factors as distillate.
   a. What is the annual consumption of natural gas? Of diesel?
   b. What is the annual emission rate for NO_x from this boiler?
   c. What is the annual emission rate for cadmium from this boiler?
   d. What is the annual potential to emit for cadmium from this boiler?

2. A diesel fired boiler operating in a serious nonattainment area is rated at 56 million Btu/hr and is seeking a permit to operate 5000 hr/year instead of its previously permitted level of 120 hr/year. The diesel has 130,000 Btu/gal energy content and 0.05% sulfur.
   a. What is the change in emissions for NO_x, SO_2, and CO for this boiler due to this change in operating hours?
   b. Does this change require a major permit modification?

3. A plant is proposed that will have two industrial boilers, each of which is rated to burn natural gas at a rate of 150,000 cubic feet per hour. The design heat release rate is 100,000 Btu/hr/ft³. The plant is to be located in an attainment area for all pollutants.
   a. What is the potential to emit for this plant for NO_x, SO_2, PM_{10}, and CO?
   b. What permitting will be needed for this plant?

4. An institutional boiler used for space heating has a heat input of 6,500,000 Btu/hr and uses natural gas as a fuel. What are its uncontrolled emissions of NO_x, VOC, SO_2, PM_{10} and CO?

5. A spray coating line is permitted to use 4000 gal/year of highly volatile lacquer for coating wood furniture. The booth has 90% efficient mat filters and uses a high volume low pressure (HVLP) spray gun that has a transfer efficiency of 65%. The lacquer has a density of 7 lb/gal, which includes 3.85 pounds/gallon of methyl isobutyl ketone (MIBK) and 1.4 lb/gal of acetone as solvents. These are the only solvents in the lacquer.
   a. What are the MIBK emissions from this source?
   b. What are the total VOC emissions from this source?
   c. What are the particulate emissions from this source?

6. A paint booth for large appliance manufacturing is being permitted. The quick-dry enamel used in the booth has a VOC content of 450 g/L, a hexavalent chromium content of 0.01%, and a density of 7 lb/gal. The booth uses electrostatic spray for application of the enamel with a transfer efficiency of 65%. It has dry mat filters for particulate control that have an 80% control efficiency. The booth is anticipated to use 1400 gallons of enamel per year.
   To control VOCs, an incinerator will be used. The incinerator has 95% control of VOCs but it emits nitrogen oxides at the rate of 300 lb/mmscf of natural

gas burned. The paint booth and incinerator are anticipated to operate 16 hours per day and 240 days per year. Natural gas use is 0.6 mmscf/day.

   a. What is the emission rate for VOC from this spray booth/incinerator system?

   b. What is the emission rate for ozone precursors from this system?

   c. What is the emission rate for hexavalent chromium from this system?

7. An air stripping tower is used to remove tetrachloroethylene (TCE) from contaminated groundwater. The concentration of TCE in the groundwater coming into the stripper is 0.370 mg per liter. The water flow through the air stripper is 2500 gal/day. The concentration of TCE in the water leaving the stripper is 0.015 mg/liter. What is the TCE emission rate from the air stripping tower?

8. A distribution center for gasoline has the following component count:

     12 pumps, 140 valves, 435 flanges, and 15 sampling locations

The concentration of benzene is 1.7% in winter blend gasoline and 2.1% in summer blend. Summer blend is used between May 1 and October 15 in a normal year. What are the annual fugitive emissions of benzene from this distribution center?

9. A source test sampled for lead in a stack gas. The sampling was for 2 hours, and during that time, 4 cubic meters of stack gas was extracted (at standard conditions). The stack gas flow rate was 800 standard cubic meters per minute. The lab sends back a result that they found 60 milligrams of lead in the sample you sent to them.

   a. What is the concentration of lead in this stack?

   b. What is the mass emission rate of lead from the stack?

REFERENCES

1. U.S. EPA, Compilation of Air Pollutant Emission Factors (AP-42), Fifth Edition, PB 86-142906, U.S. EPA, Research Triangle Park, NC. October 1995.
2. EPA Locating and Estimating Series:
3. Shareef, Gunseli S., William A., Butler, Luis A. Bravo, Margie B. Stockton, Air Emissions Species Manual, Vol. 1, Organic Compound Species Profiles EPA-450/2-88-003a, April, 1988.
4. AP-42, *op. cit.*, Section 1.3.
5. Ventura County APCD AB2588 Combustion Emission Factors, n.d.
6. API, Manual of Petroleum Measurement Standards Chapter 19 — Evaporative Loss from Fixed-Roof Tanks, API Publication 2518, American Petroleum Institute, Washington, D.C., October, 1991.
7. Ibid.
8. BACT/RACT/LAER Clearinghouse Information System (BLIS), EPA Office of Air Quality Planning and Standards (OAQPS) Technology Transfer Network (TTN), (919) 541-5742. Also on the Internet at http.//www.epa.gov/oar/ttn_bbs.html.
9. U.S. EPA. Handbook: Control Technologies for Hazardous Air Pollutants EPA/625/6-91/014, June 1991. National Air Toxics Information Clearinghouse (NATICH), 919-541-5645.

Substance	EPA Publication Number
Acrylonitrile	EPA-450/4-84-007a
Carbon Tetrachloride	EPA-450/4-84-007b
Chloroform	EPA-450/4-84-007c
Ethylene Dichloride	EPA-450/4-84-007d
Formaldehyde	EPA-450/4-84-007e
Nickel	EPA-450/4-84-007f
Chromiurn	EPA-450/4-84-007g
Manganese	EPA-450/4-84-007h
Phosgene	EPA-450/4-84-007i
Epichlorohydrin	EPA-450/4-84-007j
Vinylidene Chloride	EPA-450/4-84-007k
Ethylene Oxide	EPA-450/4-84-0071
Chlorobenzenes	EPA-450/4-84-007m
Polychlorinated Biphenyls (PCBs)	EPA-450/4-84-007n
Polycyclic Organic Matter (POM)	EPA-450/4-84-007

10. SW-846, Test Methods for Evaluating Solid Wastes, 3rd Edition, U.S. EPA, Office of Solid Waste and Emergency Response, Washington, D.C., 4 volumes.
11. U.S. EPA, Protocols for Generating Unit-Specific Emission Estimates for Equipment Leaks of VOC and VHAP. EPA-450/3-88-010, U.S. Environmental Protection Agency, Office of Air Quality Planning and Standards, Research Triangle Park, N.C., October, 1988.
12. *Environmental Reporter,* Federal Regulations 121:1564.40.
13. 40 CFR Part 60, Appendix A, Method 19.

3   BEST AVAILABLE CONTROL TECHNOLOGY

I. INTRODUCTION

Best available control technology (BACT) is defined as:

An emission limitation . . . based on the maximum degree of reduction for each pollutant . . . taking into account energy, environmental, and economic impacts and other costs . . . through application of production processes or available methods, systems, and techniques. . ."[1]

In this chapter, we will review the important elements in this definition, and describe the approach that must be used to satisfy BACT requirements for a proposed new source or modification of an existing source.

As described in Chapter 1, there is a requirement for new source review and other air quality permits that new or modified sources be built with BACT. Because BACT is required for new or modified sources, it provides a substantial incentive for developers of control technologies to show that their product provides the maximum degree of reduction of pollutant emissions available better than competing products or at a lower cost. If a new technology can be shown to reduce emissions more than currently available equipment, taking into account energy, environmental, and economic impacts and other costs, BACT requires that it will be used. This kind of requirement, which persistently forces improved technologies on sources of air pollutant emissions, is termed *technology forcing*.

BACT is intended to include end-of-pipe controls, such as scrubbers or baghouses; process changes, such as changing from a solvent- to a water-based coating; and operations, such as inspection and maintenance of valves and flanges to reduce fugitive emissions. The intent is that there is no technology feasible to use that would result in lower emissions.

However, the definition also provides that the evaluation of BACT for a source will be on a case-by-case basis in which the characteristics and constraints of the source will be considered. The definition calls for evaluation of energy, environmental, and economic impacts and other costs that are specific to the source being considered. If these impacts are too great for a pollution control technology applied to a specific source, that technology is not considered BACT *for that case*. In another case, different conditions can lead to a different conclusion.

In the following sections, several additional definitions will be offered for terms that are similar to BACT but apply to special situations. Next, strategies for use in BACT determination for a proposed project are discussed. Then the "top-down" approach to determine BACT is described. This is the approach required by most permitting agencies. Finally, several examples are given for determining BACT.

II. CONTROL TECHNOLOGY REQUIREMENT DEFINITIONS

In addition to BACT, there are several other technology requirement terms that may be encountered in air quality permitting. All of them are conceptually similar to BACT — they define emission limits applicable to different source types, different pollutants, and/or different attainment status. They are determined in similar fashions, and in some cases, they may be more stringent. Several of them are discussed below.

THERE ARE SEVERAL DIFFERENT CONTROL TECHNOLOGY REQUIREMENTS

- BACT - Best available control technology
- LAER - Lowest achievable emission rate
- RACT - Reasonably available control technology
- MACT - Maximum achievable control technology
- GACT - Generally available control technology
- Others include T-BACT, BARCT, RACM, and BACM

A. Lowest Achievable Emission Rate (LAER)

LAER is the control technology that is required for nonattainment pollutant sources. If an area is nonattainment for ozone, LAER will be required for new or modified sources that emit the ozone precursors, nitrogen oxides (NO_x) and volatile organic compounds (VOC). LAER is defined as follows:

For any source, that rate of emissions which reflects —

(A) the most stringent emission limitation which is contained in the implementation plan of any state for such class or category of source, unless the owner or operator of the proposed source demonstrates that such limitations are not achievable, or

(B) the most stringent emission limitation which is achieved in practice by such class or category of source, whichever is more stringent"[2]

The stringency of LAER is its absence of extenuating circumstances such as those contained in the definition of BACT. Although there is still a provision "unless the owner or operator of the proposed source demonstrates that such limitations are not achievable," it only applies to controls identified in implementation plans, and any control "achieved in practice" must be adopted. The consequence of this language is that economic feasibility is not included, and a proposed source must install the most stringent control that has been "achieved in practice." As applied, there has been some debate about whether "achieved in practice" counts if the "practice" is outside the United States.

B. California BACT

Definitions for BACT that conform to a version of the following format have been adopted by 24 California air pollution control districts:

BACT means for any (source, stationary source, emission unit) the most stringent of:

(a) The most {stringent, effective} {emission limitation, emission control} [or control technique] which the EPA {certifies, states} is contained in the implementation plan of any state approved under the Clean Air Act for {such category or class of source, the type of equipment comprising such a source}, unless the applicant demonstrates to the satisfaction of the Air Pollution Control Officer (APCO) that such limitation is not achievable.

(b) The most effective control device, technique, or emission limit which has been achieved in practice for such category or class of source.

(c) Any other emission control technique [alternative basic equipment, different fuel or process] found [after public hearing] by the APCO to be technologically feasible and cost effective for such class or category of sources [or for a specific source].

Under no circumstances shall BACT be determined to be less stringent than the emission control required by any applicable provision of District, state, or federal laws or regulations, unless the applicant demonstrates to the satisfaction of the APCO that such limitations are not achievable.

Part (c) of this definition is more stringent than federal LAER, because it requires consideration of control technologies that are technologically feasible and cost effective, but not yet achieved in practice on that particular source type.[3]

C. Reasonably Available Control Technology (RACT)

RACT is the term that is applied when a State Implementation Plan (SIP) calls for reductions in emissions of existing sources in nonattainment areas in order to progress toward attainment. Typically, RACT is implemented as a prohibitionary rule applicable to nonattainment areas of a state or region. An example would be a limitation on NO_x emissions from a boiler larger than 50,000 Btu/hr design capacity. An RACT rule would require that all boilers above this size have NO_x emissions less than 0.5 lb/MMBtu of heat input, and that existing boilers would have 3 years after the date of promulgation of the rule to comply. RACT requirements are never more stringent than BACT, and are usually less stringent because they must be retrofitted to existing equipment. Since RACT applies to existing sources, it is not a requirement which must be addressed for new sources. However, it may become an issue when a new source is seeking offsets from an existing source being taken out of service. This is discussed further in Chapter 4.

D. Best Available Retrofit Control Technology (BARCT)

BARCT is a California term that refers to retrofit controls for existing sources in nonattainment with California ambient air quality standards. It is defined in the California Clean Air Act. It is difficult to distinguish between RACT and BARCT. Both apply to existing sources. Both are only applicable in nonattainment areas. BARCT is sometimes more stringent than RACT for a given source type. The California Air Resources Board (CARB) periodically publishes technical memoranda identifying BARCT for particular source types. As with RACT, BARCT is implemented through prohibitionary rules within each of the 34 California air pollution control districts.

E. Maximum Achievable Control Technology (MACT)

MACT is the control level for hazardous air pollutant emission sources required under the Clean Air Act Amendments of 1990. The definition in the Act is:

> . . . the maximum degree of reduction in emissions of hazardous air pollutants subject to [Section 112 of the Act] (including a prohibition on emissions, where achievable) that the Administrator, taking into consideration the cost of achieving such emission reductions, and any non-air quality health and environmental impacts and energy requirements, determines is achievable for new or existing sources in the category or subcategory to which such emission standard applies . . .[4]

MACT is discussed in more detail in Chapter 6, since it addresses hazardous air pollutant control. It also is distanced in its definition from LAER, as it is not intended to ignore economic considerations.

F. Generally Available Control Technology (GACT)

GACT, like MACT, is a term that addresses hazardous air pollutant control. GACT is the control technology standard that is applied to non-major hazardous air pollutant sources. Since these sources emit smaller amounts of hazardous air pollutants, GACT may be less stringent than MACT, although for some hazardous air pollutant source categories for which MACT standards have been promulgated, the GACT standard has been the same. GACT is discussed further in Chapter 6.

G. Toxic Best Available Control Technology (T-BACT)

T-BACT was coined by South Coast Air Quality Management District (SCAQMD) in California to identify technologies that were suitable for control of toxic air pollutants. The term was adopted by several other Districts in California, and applies to new sources of toxic air pollutants in those jurisdictions that have estimated health risk above a threshold value of 1 in 100,000 cases of cancer due to those emissions. In practice, T-BACT control technologies are the same as technologies for control of PM_{10} and VOC in almost every circumstance, so this term has not become widespread. For a more complete discussion of toxic air pollutants and health risk, see Chapter 6.

H. Reasonably Available Control Measures (RACM) and Best Available Control Measures (BACM)

RACM and BACM are control measure guidelines developed by EPA for inclusion in State Implementation Plans. Although these terms can be applied to any pollutant, they have been most commonly applied to PM_{10} sources. The terms RACM and BACM address moderate and serious PM_{10} nonattainment areas respectively. RACM and BACM technical guidance was mandated under Section 190 of the Clean Air Act Amendments of 1990 (CAAA) for three major PM_{10} source categories: urban fugitive dust,[5] residential wood combustion, and prescribed silvicultural and agricultural burning. These measures were developed to use as prohibitory rules, similar to those implementing RACT, aimed at reducing PM_{10} emissions with the goal of reaching attainment for those PM_{10} nonattainment areas. Measures identified as RACM and BACM will almost certainly be on the list of alternative technologies to be considered as possible BACT for sources of PM_{10} .

III. BACT SELECTION STRATEGY

Most jurisdictions require that BACT be used in all new or modified sources. The regulations seldom specify how to determine BACT beyond defining the term. In Section IV below, the "top-down" approach developed by EPA for formal determination of BACT is described. The "top-down" process is thorough, sequential, and complete. However, it can be expensive to conduct and may result in the applicant being required to use a technology not desired. This dilemma can sometimes be avoided.

In most cases, both the applicant and the permitting agency agree on what is the most stringent technology for the source being permitted. If the applicant decides to use that control technology, a "top-down" BACT process is not needed, and accepting that most stringent technology early in the permitting process can greatly speed up other approval steps.

In other cases, the number of technologies reviewed in the "top-down" process can be limited by prior agreement with the permitting agency. The number may be limited by citing BACT analyses that have been done for similar processes in similar settings and having the agency accept these analyses as applicable.

Sometimes it may be worthwhile to prepare a BACT protocol similar to the AQIA protocol described in Chapter 5. In a BACT protocol, the project description can be summarized, assumptions can be stated, and technologies that will be considered as part of the BACT analysis identified. This document can clearly set the boundaries of the BACT analysis and avoid surprises later in the process. Technical feasibility is usually used as the basis for including technologies, and infeasible technologies can be excluded prior to the "official" start of the analysis. After a BACT protocol has been submitted, a telephone call is often adequate to agree on the boundaries of the analysis.

There are many instances when only a few (often as few as two) technologies are considered in the BACT analysis, and in those situations, the permitting agency may accept a much less formal analysis than the "top-down" approach. Brevity usually makes the analysis easier to understand and allows the opportunity to make important points clearly without unnecessary boilerplate.

IV. TOP-DOWN BACT ANALYSIS

In 1987, EPA issued a memorandum that described a "top-down" approach to determining what constitutes BACT for a particular source.[6] The memorandum is to be followed up with regulatory language in 1996. The fundamental guide of the "top-down" approach is that the examination of what control technology constitutes BACT will start with the most stringent technology — that is, the one with the greatest emission reduction. The applicant must demonstrate that this most stringent technology will not work in order to reject it. The basis for it not working can be technical infeasibility, or because of energy, environmental, economic or other cost impacts that are unacceptable. The assumption starting the analysis is that the most stringent technology will work, and should be used, unless proven otherwise. The process of disproving feasibility of successively less stringent technologies continues until a control technology is found that will work, and that technology becomes BACT for that particular source. In the following subsections, the steps required in a "top-down" BACT analysis are described.

A. Identify Technologies

The first step in the "top-down" process is to identify the control technologies that will be considered in the analysis. Historically, intense arguments have arisen over this question, mostly related to whether a technology proposed by a regulatory agency for inclusion on the "top-down" list was available. When addressing this issue, the California BACT provision to include "any other emission control technique found by the APCO to be technologically feasible . . ." can give rise to suggestions to consider many different technologies that may not yet be commercially available.

The availability criteria used by SCAQMD contains most of the elements used elsewhere. The SCAQMD list includes the following:[7]

1. The technology is commercially available.
2. The technology is reliable. There has been at least 12 months of demonstrated performance in comparable field conditions.
3. The technology is effective as demonstrated by a source test or other performance documentation.
4. The technology has been achieved in practice or is technologically feasible and cost effective.

Lists have been compiled of control technologies for different equipment or processes.[8-10] Descriptions of control technologies are also available.[11-12] These references, which contain very similar information, are adequate to provide a list of alternatives that must be considered in the "top-down" analysis for most jurisdictions, and they are updated periodically. Some of the lists include technologies considered LAER as the most stringent technology listed. The logic for including LAER is that this is the most stringent technology, and the applicant must disprove its use in the "top-down" process.

Process changes that result in lower emissions can also be considered BACT. In some instances, process changes such as changing from a solvent- to a water-based coating can result in much lower emissions and fewer hazardous waste products.

There is opportunity for a great deal of judgment in the selection of technologies for the "top-down" list, and that judgment is exercised differently by agencies in different states or regions. Including technologies on the initial "top-down" list that have little or no likelihood of being technically feasible for a particular source increases the amount of work that must be done to exclude that technology from the process.

The term *available* with respect to control technologies refers to commercial availability. Can you purchase this technology, and has it been demonstrated in a commercial setting? The commercial setting means that it must have been part of a commercial process, and not a research effort, pilot test, or demonstration.

There is sometimes disagreement about whether a technology that has been used commercially in a foreign country is considered "available" in the United States. Often there is not good information available about foreign "commercial technologies" to determine whether the use was a commercial setting and not subsidized by the technology vendor or an agreement that could reimburse the user if the technology were adopted as BACT in the United States. Usually if there are several "apparently commercial" applications in other countries, the assumption of commercial availability is made for the purposes of including the technology in a BACT list. However, of the references cited above for lists of BACT technologies, foreign uses have not been included.

B. Determine Technical Feasibility

Each of the technologies identified must be technically feasible for the source being permitted. The technology must be able to be used in the particular circumstances of the source being permitted. For example, a technology to reduce

NO_x from combustion turbines is to use water injection into the combustion chambers. However, for a gas pipeline compressor station in a remote, arid area where water is not available, water injection is not considered technically feasible. Another example would be the use of floating roof organic liquid storage tanks. This technology is considered technically feasible in most areas, but in extremely cold, icy environments such as on the North Slope, the sliding seal on floating roof tanks tends to freeze in place, preventing the roof from moving. Because of this, floating roof tanks are considered technically infeasible in Arctic locations.

In some instances, prior agreement can be negotiated with the agency to exclude technologies that have been shown to be infeasible in recent similar analyses. Such agreements should always be confirmed in writing to assure that a notice of deficiency is not issued for the subsequent application.

There can be some uncertainty about whether a technology is technically infeasible for a particular application, or whether it would just be considered too costly. If the issue were cost, it would need to be addressed in the cost evaluation later in the "top-down" process.

For most of the arguments of technical infeasibility, an expensive approach can be imagined to make it feasible. For example, for the NO_x control by water injection case, drilling for water could be done, and at some depth, water could probably be found. The cost of obtaining water would be extremely high but possible! The strategy to use in the BACT analysis should be to propose a limitation as to technical infeasibility. Then the additional effort needed to estimate costs is not necessary — the technology is off the list. If the technical infeasibility argument is rejected, additional analysis will be needed to demonstrate either that the technology is too expensive or that the technology could be defined as BACT. Note that in applications where LAER is required, the cost argument is not usable, so proposing technical infeasibility may be the only choice.

C. Rank Technically Feasible Alternatives

Once a "short list" of technically feasible control alternatives has been developed, the alternatives are ranked from the most to least effective in reducing emissions from the proposed source. Often it provides clarification to express control level in more than one way, such as percent control and concentration of pollutants in an exhaust stream. The most useful common measure of control effectiveness is the total emissions per year from the source, using the control technology identified. Technologies are listed "top-down," starting with the most effective technology that is technically feasible.

This short list should be presented to the permitting agency and agreed upon before continuing the BACT analysis. Once the list of technologies has been agreed to between the agency and the applicant, the remaining steps in the "top-down" process take place one technology at a time, starting with the "top" or most effective technology. The intent of the next steps is to determine whether there are reasons for rejecting the application of this "top" technology to the process or equipment being permitted.

D. Evaluate Impacts of Technology

The next step in the "top-down" process is to determine whether there are energy, environmental, or cost impacts from the use of this top technology that would render it infeasible.

1. Energy Impacts

Energy impacts include the energy consumed by the control technology or a reduction in energy produced, if an energy-production process is permitted. One example of an energy impact would be the need to add heat to a low-temperature flue gas to increase its temperature to the place where a selective catalytic reduction system that operates at a minimum temperature of about 600°F could be used for NO_x control. Quantitative thresholds for energy impacts have not been published, so an energy impact rejection of a top technology is discretionary.

One possible basis for rejection of a technology based on energy impacts would be *reductio ad absurdum*. Many years ago, an analysis was done of possible technologies for reducing CO_2 emissions from an electric utility power plant. At that time, the only process for reducing CO_2 was a high-pressure amine scrubber. The analysis of this alternative indicated that in order to scrub the CO_2 from the power plant exhaust, more energy would be required than was produced by the power plant. Such a technology could be eliminated based on energy impacts.

2. Environmental Impacts

If use of a proposed control technology results in other environmental impacts, the technology may be rejected because of other environmental impacts. For example, the use of the technology might result in the production of hazardous wastes, the creation of other air pollutants than those being controlled with the technology, excess water use in an area with inadequate water supply, generation of large quantities of solid wastes or solid wastes that are difficult to dispose of, or even destruction of critical habitat if the footprint of a control technology installation extended beyond currently disturbed areas.

3. Cost Effectiveness

The cost effectiveness of a control technology is measured in dollars per ton of pollutant removed. It is determined by first calculating the rate of pollutant removal (tons/year). This rate is the difference between the emissions before the control is installed and the rate after control. Then the annual cost of the technology (dollars/year) is determined. Cost effectiveness is the ratio of the annual cost of the technology to the rate of pollutant removal (dollars/ton). The control efficiency and rate of pollutant removal were discussed earlier. In this section, calculating the annual cost will be discussed. A form for estimating capital costs, with default values for different cost elements, is shown in Table 3-1. A form addressing annual cost estimates is included as Table 3-2. The default values should be used only when project specific cost data are not available.

Table 3-1   Capital Cost Estimating Table

Direct costs	
Purchased equipment	
Basic equipment and auxiliaries	A
Freight	<u>0.05A</u>
Total purchased equipment cost	B
Direct installation costs	
Foundations and supports	0.08B
Erection and handling	0.14B
Electrical	0.04B
Piping	0.02B
Painting	0.01B
Insulation	0.01B
Building and site preparation as necessary	
Total direct installation cost	0.30 + Building and site preparation
Total direct cost	1.30 + Building and site preparation

Indirect costs	
Engineering and supervision	0.10B
Construction and field expenses	0.05B
Construction fee	0.10B
Start-up	0.02B
Contingency	0.03B
Total indirect cost	0.30B
Total capital cost	1.60B + Building and site preparation

Source: W.M. Vatavuk, OAQPS Control Cost Manual, (Fourth Edition), EPA/450/3-90/006, January, 1990

Table 3-2   Annual Cost Estimating Table

Direct costs	
Operating labor	0.5 hr/shift $35/labor-hour
Supervisory labor	15% of operating labor
Maintenance labor	0.5 hr/shift $35/labor-hour
Maintenance materials	100% of maintenance labor
Replacement parts	Catalyst replacement every three years, catalyst cost approximately 50% of purchased equipment cost
Utilities	
Electricity	$0.03/kWh
Ammonia	$200/ton

Indirect costs	
Overhead	60% of sum of operating, supervisory, and maintenance labor plus maintenance materials
Administrative charges	2% of total capital cost
Property tax	1% of total capital cost
Insurance	1% of total capital cost
Capital recovery cost	Product of capital recovery factor and total capital cost
Annual cost	Sum of direct and indirect annual costs
Cost-effectiveness	Annual cost divided by emissions controlled

Source: W.M. Vatavuk, OAQPS Control Cost Manual, (Fourth Edition), EPA/450/3-90/006, January, 1990

The annual cost of a control technology is the sum of the annualized installed capital cost and the annual operating and maintenance costs. The installed capital cost, according to EPA guidelines, should be a plus or minus 30% accuracy.[13] Information for estimating pollution control equipment costs and standard assumptions used in control cost estimating is available,[14] but confirmation and updates should be sought from other sources. These include EPA Control Technology Guideline documents, control cost data from trade publications, and vendor cost estimates. Usually at the time when the BACT analysis is being done, detailed design costs or bids for construction are not available, and so construction cost estimates are not expected to be used in the cost effectiveness calculation.

The installed capital cost should be specific to the site and equipment being permitted. In particular, remote sites may have very high installation costs that could affect the determination of overall cost effectiveness of a technology. Also, extraordinary costs unique to the proposed application must be justified if they differ from standard cost assumptions. Indirect costs as well as working capital requirements should be included if appropriate.

One of the substantial challenges associated with estimating capital costs is deciding what equipment to include in the cost calculation. Any cost that must be incurred to make the control device work is justifiably included in the capital cost calculation. This might include electrical systems, waste handling systems, and automated control equipment. Including in the cost equipment that would be needed whether or not the control equipment was present would not be justified. For example, the cost of a stack could not be included as a capital cost associated with a control technology, but it might be argued that increased stack height needed because of a temperature drop in the control device is an expense associated with that control.

The installed capital cost must be amortized over the life of the project to determine the annual cost component due to capital costs. The capital recovery factor (CRF) is multiplied by the capital cost to obtain the annualized cost. The capital recovery factor is given by:

$$CRF = \frac{i(1+i)^n}{\left[(1+i)^n - 1\right]} \tag{3-1}$$

where i = interest rate and n = lifetime of control system.

Historically, an annual interest rate of 10% has been used in cost calculations. If interest rates for industrial projects are substantially different than 10% at the time of the application, the current rate may need to be used.

The project lifetime is usually included in the project description. Lifetimes for industrial equipment are assumed to be 10 years, since a major overhaul is usually required within that time. A lifetime of 10 years is the nominal value recommended by regulatory agencies unless a different value can be justified. For an interest rate of 10% (i = 0.1) and a lifetime of 10 years (n = 10), the capital recovery factor equals 0.163 per year.

All operating and maintenance costs (O&M) should be included when determining annualized costs. O&M costs should include fuel required, routine maintenance costs, insurance, taxes, utilities, replacement parts, raw materials purchased, waste material disposal costs, overhead, and general and administrative (G&A) costs.

4. Maximum Cost Effectiveness Criteria

The basis for a cost decision on a control technology being designated as BACT is whether it is too expensive. Thus a criterion is needed. Each permitting agency can decide what their criteria for maximum costs are. Usually these criteria are expressed in terms of cost effectiveness (dollars/ton removed), as calculated above.

In California starting in 1987, the criteria used were those that had been adopted by SCAQMD. In 1995, SCAQMD updated their criteria based on costs of control for different technologies. Both the 1987 and 1995 SCAQMD criteria are shown in Table 3-3.

Table 3-3   SCAQMD BACT Maximum Cost Effectiveness[15]

Pollutant	1987 Criteria ($/ton removed)	1995 Criteria ($/ton removed)
ROG	$17,500	$18,000
NO_x	$24,500	$17,000
SOx	$18,300	$9,000
TSP(PM_{10})	$5,300	$4,000
CO	—	$350

Source: South Coast Air Quality Management District, "Draft Best Available Control Technology Guidelines, Part A: User's Guide and Part B: BACT Determinations, September 11, 1995.

The selection of maximum cost effectiveness criteria is somewhat arbitrary. Often permitting agencies are open to arguments about the appropriateness of criteria selected, and particularly in cases where large capital investments are necessary, the agency may be open to rejecting an expensive technology even if it has good cost effectiveness. If the cost effectiveness of the most efficient technology is significantly higher than the next most effective, this can be used as an argument for rejecting that technology.

V.   BACT ANALYSIS EXAMPLE

In the following section, an example is provided of a BACT analysis for a new source. Although the specific data in the analysis may not be applicable to another application, the approach should be consistent.

A. Boiler NO$_x$ Control

1. Project Description

A 120 million Btu/hr natural gas-fired boiler is planned for installation to provide steam for startup and as a backup for a gas-fired combustion turbine cogeneration facility serving a rendering plant. The planned capacity factor for this boiler is 18% or 1600 hr/yr.

2. Identify Technologies

The technologies that can be considered for the "top-down" BACT analysis are shown in Table 3-4. The list was adapted from references 5 to 8 and vendor literature.

Table 3-4   NO$_x$ Control Technology Options

Control technology	Control efficiency
Selective catalytic reduction plus low NO$_x$ burner plus flue gas recirculation	90%
Selective noncatalytic reduction plus low NO$_x$ burner plus flue gas recirculation	85%
Low NO$_x$ burner plus flue gas recirculation	80%
Flue gas recirculation	70%
Low excess air	30%

Source: Adapted from references 5-8 and vendor literature.

3. Determine Technical Feasibility

All of the technologies identified have been demonstrated in numerous applications over several years. For this application, there were no constraints that would prohibit use of these technologies.

4. Rank Technically Feasible Alternatives

Table 3-5 presents the technically feasible alternatives with the resulting emissions assuming this control. The list is presented in "top-down" fashion from the most effective to the least effective control.

Table 3-5   Emission Estimates with NO$_x$ Control Technology Options

Control technology	Emissions (tons/year)
Baseline emissions with new source performance standards	9.6
Selective catalytic reduction plus low NO$_x$ burner plus Flue gas recirculation	0.96
Selective noncatalytic reduction plus low NO$_x$ burner plus flue gas recirculation	1.4
Low NO$_x$ burner plus flue gas recirculation	1.9
Flue gas recirculation	2.9
Low excess air	6.7

The process continues from this point analyzing technologies sequentially from the topdown. Each technology, starting with the most effective, is examined to determine whether there are reasons it will not be feasible for this application. Consequently, the analysis continues looking first at Selective Catalytic Reduction (SCR) plus low NO_x burner plus Flue Gas Recirculation (FGR). This combination of control technologies can achieve 9 ppm (at 3% O2) and emissions of 0.96 tons/year of NO_x.

5. Evaluate Impacts of Technology

Energy. Because the SCR system includes a catalyst bed, there is a pressure drop through the device, and a larger induced draft (ID) fan is required than for a boiler without an SCR. The energy penalty for this fan is 5% of the boiler output. The FGR system also requires a fan to recirculate the gases. The energy penalty for this fan is about 1% of boiler output.

Environmental Impacts. The SCR system contains a precious metal (vanadium or titanium based) catalyst that must be replaced every 3 years. The spent catalyst must be disposed of as a hazardous waste, creating a potential environmental impact. Since this boiler will function as a backup, it will be required to come on line quickly. During the period before the catalyst bed is up to temperature, there will be excess NO_x emissions from an SCR-based control system.

Cost Effectiveness. The capital cost for an SCR system for a boiler of this size would be $600,000, virtually the same cost as the package boiler it would control. In addition to normal operating costs, extraordinary operating costs would include ammonia and increased fuel use to compensate for the energy loss of 6% discussed above. The annual cost for the boiler with SCR and the cost effectiveness are shown in Table 3-6. The cost effectiveness of SCR for this application is $30,008 per ton of NO_x removed, so based on the criteria given in Table 3-3, SCR would not be considered cost effective, and hence would not be BACT.

Table 3-6   Capital Cost Components for SCR Control of Example Boiler

Direct costs	
Purchased equipment	
Basic equipment and auxiliaries	$600,000
Freight	30,000
Total purchased equipment cost	$630,000
Direct installation costs	
Foundations and supports	$ 50,400
Erection and handling	88,200
Electrical	25,200
Piping	12,600
Painting	6,300
Insulation	6,300
Building and site preparation as necessary	
Total direct installation cost	$189,000
Total direct cost	$819,000 + Building and Site Preparation

Table 3-6   Capital Cost Components for SCR Control of Example Boiler (Continued)

Indirect costs	
Engineering and supervision	$ 63,000
Construction and field expenses	31,500
Construction fee	63,000
Start-up	12,600
Contingency	18,900
Total indirect cost	$189,000
Total capital cost	$1,008,000

Direct costs	
Operating labor (3 shifts/day, 200 days/year)	$ 10,500
Supervisory labor	$ 1,575
Maintenance labor	$ 10,500
Maintenance materials	$ 10,500
Replacement parts — catalyst replacement every three years	$105,000
Utilities	
Electricity	0
Ammonia (1 lb. ammonia/lb NO_x removed)	$ 1,728

Indirect costs	
Overhead	$ 19,845
Administrative charges	$ 20,160
Property tax	$ 10,080
Insurance	$ 10,080
Capital recovery cost (10% interest for 10 years)	$164,304
Annual cost	$259,272
Cost-effectiveness	$30,008/ton removed

Source: L. P. Nelson, Radian International LLC

REFERENCES

1. This definition is abbreviated from that found in 40 Code of Federal Regulations (CFR) 52.21(j) related to the BACT requirement for PSD permitting. Also see the Clean Air Act, §165(a)(4) and 169(3).
2. 40 CFR 51.165(a)(1)(xiii). Also see the Clean Air Act §171(3).
3. South Coast Air Quality Management District, "Draft Best Available Control Technology Methodology Report," September, 1995, p. 20.
4. Clean Air Act Section 112(d)(2).
5. U.S. EPA, "Fugitive Dust Background Document and Technical Information Document for Best Available Control Measures," Office of Air Quality Planning and Standards, Research Triangle Park, N.C. EPA-450/2-92-004, September, 1992.
6. McCutchen, Gary, "Draft New Source Review Workshop Manual, Prevention of Significant Deterioration and Nonattainment Permitting," U.S. EPA Office of Air Quality Planning and Standards, October, 1990, p. B-35. To be promulgated as regulations, 1996.

7. South Coast Air Quality Management District, "Draft Best Available Control Technology Guidelines, Part A: User's Guide and Part B: BACT Determinations, September 11, 1995.

8. *Ibid.*

9. California Air Pollution Control Officers Association, "A Compilation of California BACT Determinations Received by the CAPCOA BACT Clearinghouse (Second Edition)," California Air Resources Board Stationary Source Division, November, 1993.

10. BACT/RACT/LAER Clearinghouse Information System (BLIS), EPA Office of Air Quality Planning and Standards (OAQPS) Technology Transfer Network (TTN), (919)541-5742. An article describing the process of accessing BLIS can be found in *The Air Pollution Consultant,* November/December, 1994, Elsevier Science Inc., New York, p. 4.18.

11. Theodore, Louis and Anthony J. Buonicore, *Air Pollution Control Equipment,* CRC Press, Boca Raton, FL, Vol. 1: *Particulates,* 184 pp., and Vol. 2: *Gases,* 160 pp., 1988.

12. Buonicore, Anthony J. and Wayne T. Davis, Editors, *Air Pollution Engineering Manual,* Van Nostrand Reinhold, New York, 1992.

13. McCutchen, *op. cit.*

14. Vatavuk, William M., *Estimating the Costs of Air Pollution Control,* Lewis Publishers, Boca Raton, FL, 300 pp., 1990.

15. South Coast Air Quality Management District, *op. cit.,* p. 64.

4   EMISSION REDUCTIONS AND OFFSETS

I.   INTRODUCTION

In areas that are nonattainment — those that do not meet the ambient air quality standard for one or more criteria pollutants — there is a requirement that any proposed significant increase in air pollutant emissions must be *offset* by a corresponding decrease in emissions from an existing source. The intent of the requirement is that a nonattainment area cannot sustain further increases in pollutant emissions, and emission reductions are usually required to be greater than the proposed increases from a new or modified source. The ratio of required reductions to proposed increases is termed the offset ratio.

Emission reductions that are certified are termed emission reduction credits (ERC). For emission reductions to be certified as ERC, they must have five characteristics shown below. The requirements for certification are discussed below.

REQUIRED CHARACTERISTICS OF EMISSION REDUCTION CREDITS

- Real
- Quantifiable
- Surplus
- Permanent
- Enforceable

Real. The only emission reductions that can count as ERCs must be actual emissions, not permitted levels.

Quantifiable. ERCs claims must be based on secure documentation, either through source tests, mass balances, or other means acceptable to the permitting agency. See Chapter 2 for a more extensive discussion of emission estimating techniques.

Surplus. ERCs must be emission reductions beyond levels required by regulations such as RACT.

Permanent. Any reductions claimed as ERCs must be permanent reductions. An exception to this is mobile source ERCs discussed below.

Enforceable. The emission reductions must be secured by permit modifications, relinquishing permits for shutdown sources, or other legally enforceable means.

If a facility desiring to increase its emissions does not have the potential for offsets by reducing emissions from existing sources, it must locate emission reduction credits from other sources. Usually, it is necessary to purchase these credits by paying for the pollution control equipment needed to effect any reduction, and sometimes paying a premium that depends on the market demand for such reductions. In some areas of the country, the demand for emission reduction credits to be used as offsets is great enough that an active market exists. However, in most areas, the market is not well developed, and emission reduction credit transactions are infrequent.

The steps involved in identifying the need for emission offsets, the quantities required, potential sources of emission reduction credits, characteristics that must be satisfied for certification, emission reduction credit prices, and other requirements are discussed below.

II. NEED FOR EMISSION OFFSETS

As discussed in Chapter 1, one of the thresholds that is important in air quality permitting is the threshold above which emission reduction offsets are required. Prior to the 1990 Clean Air Act Amendments, this threshold was usually 250 lb/day for nonattainment criteria pollutants other than carbon monoxide (CO), and 550 lb/day for CO. The 1990 Amendments required that offsets be provided for significant emission increases, and the definition of "significant" varied depending upon the severity of nonattainment. In addition, reductions in the offset threshold have been decreasing as air agencies have attempted to reduce total emissions and improve air quality. In many California nonattainment areas, permitting agencies are reducing the offset threshold to near zero. That is, any new source of air pollutants that must obtain an air permit will be required to obtain emission reductions for its proposed emissions. In some jurisdictions, emission reduction offsets are required for all pollutants, not just those that are nonattainment. This is not mandated under the Federal law, but state and local jurisdictions have the option of imposing more stringent standards than the federal law requires, and emission reduction offset requirements can be imposed for all criteria pollutants using this option.

Another offset requirement concerns which pollutants must be offset. As identified above, pollutants that are nonattainment and that are emitted above the offset threshold must be offset. Additionally, precursors to nonattainment pollutants may need to be offset.

Ozone is a secondary pollutant not emitted directly from sources. Therefore, offsets must be provided for ozone precursors in ozone nonattainment areas. The ozone precursors are nitrogen oxides, NO_x, and volatile organic compounds (VOC) that are reactive in the atmosphere.*

Particulate matter less than 10 µm (PM_{10}) may also be considered a secondary pollutant as well as a primary pollutant. Secondary PM_{10} is formed from oxidation of NO_x and sulfur dioxide to nitrates and sulfates. These chemical reactions can take considerable time to occur. Averaged over 24 hr, typically 4% of NO_x is converted to nitrates every hour after its emission into the atmosphere.[1] Because of the long reaction time involved, PM_{10} formed from NO_x and sulfur dioxide may be transported by the wind outside the nonattainment area before creating particulates. Because of this, permitting agencies differ on whether NO_x and SO_2 must be offset because of PM_{10} nonattainment.

Seasonal sources may also require offsets, or be considered as opportunities for emission reduction credits that could be used as offsets. Examples of seasonal sources would be emissions from food production or processing, such as sugar refineries or fruit or vegetable canning plants. Such plants usually operate during and after harvest seasons for periods as short as 6 weeks to as long as 6 or 7 months. Agencies regulating these seasonal sources have established quarterly seasons, which must be consistent between the new emissions and the emission reductions. For a continuous (year-round) source to utilize emission reductions from seasonal sources, seasonal sources must be reduced during all four seasons. The scientific logic of this approach can be questioned because the "cooking" time for ozone production is from 1 to 3 days, not a quarter of a year. However, uncertainties in when seasons start and end for seasonal sources (dictated by harvests, weather, and seasonal economic demand) argue against rigid dictates of day-by-day correspondence between emission reductions and new emissions.

An emerging regulatory requirement is to require offsets for toxic air pollutants. Regulations have been proposed that would require toxic air pollutant emission reductions adequate to offset toxic air pollutant increases from a new or modified source. The offset ratio between emission reductions and new emissions would be established to provide no net increase in health risk or burden from the combination of reductions and the new emissions. That is, the total number of cancer cases predicted in population affected by the new or modified source would be fewer than the reduction in cases predicted by reducing emissions from existing sources. The accuracy of estimating health risk is currently limited. Consequently, the feasibility of this toxic offset regulation is shaky, and its application is currently limited to toxic emission increases and reductions from the same facility, so that determining cancer burden does not involve two separate populations. Although it would be conceivable to calculate the increase and reduction in risk for two separate populations, the social acceptability of increasing the risk of cancer to one population while decreasing it for a different population is unlikely.

As noted in Chapter 1, permitting rules for your jurisdiction should be consulted to determine what pollutant emissions must be offset.

* Organic compounds that are considered reactive in the atmosphere are discussed in Chapter 1.

III. OFFSET QUANTITIES REQUIRED

The federal Clean Air Act requires that areas of the country that are nonattainment for any of the criteria pollutants must develop a plan by which emissions can be reduced and attainment achieved. Emission reduction offsets for new or modified sources with emission increases are required to be part of that plan under the provisions of the 1990 Amendments. Additionally, depending on the degree of nonattainment, emission reduction offsets must be greater than the proposed increases in emissions by an offset ratio shown in Table 4-1. As nonattainment becomes more severe, the quantity of emission reduction credits that must be obtained is larger for each emission increase.

Table 4-1   Offset Ratios for Different
             Categories of Nonattainment

Nonattainment category	Offset ratio
Marginal	1.1 to 1
Moderate	1.15 to 1
Serious	1.2 to 1
Severe	1.3 to 1
Extreme	1.5 to 1

Source: Clean Air Act Amendments of 1990, Title I

Again, local or state jurisdictions may impose even greater ratios than those mandated under the act. In some jurisdictions, the offset ratio increases as the distance between a new source as an emission reduction increases. This rationale may have merit for nonattainment primary pollutants, such as CO, but because of the length of time required for the atmospheric chemical reactions in ozone formation, substantial mixing will have occurred in the atmosphere, and the location of the source of precursor pollutants is no longer important.

Some jurisdictions, recognizing the regional character of ozone pollution — that the location of ozone precursors within an air basin is not important, but the quantity emitted is — have established offset ratios for ozone precursors that increase with the magnitude of emissions of a proposed source rather than the distance between the proposed source and the location of emission reductions.

Emission reduction credits from existing sources are currently determined based on the difference between emissions after a reduction and the emissions from the source assuming it has been implemented with Reasonably Available Control Technology (RACT).* In nonattainment areas, existing sources are required to install RACT to reduce emissions so that attainment can be achieved. Even if a source has not yet installed RACT, usually because RACT rules allow several years for implementation, the assumption in calculating an ERC starts with that emission level.

Another method for determining offset ratios is through the use of modeling to demonstrate a net air quality benefit. This method requires that the applicant demonstrate through dispersion modeling (see Chapter 5) that an emission reduc-

* RACT is defined and discussed in Chapter 2.

tion combined with a proposed increase would result in a "net air quality benefit" for all affected receptors. One problem with this approach is that it is easier to demonstrate a "net air quality benefit" if the existing source (whose emissions are being reduced) has very poor dispersion characteristics. An existing source with a low stack height or flue gas temperature does not disperse well, and consequently can be used to obtain credits at a much greater distance than emission reductions from a well-dispersed source.

The most common offset ratio basis currently used is based upon fixed ratios that increase with the distance between the emission reduction and the new or modified source. Ratios vary between 1:1 for a new source and emission reduction in the same facility to 2:1 or greater for distances above 50 miles between the new source and the emission reduction.

Offset ratios established by proximity between a new source and an emission reduction are in addition to those set by the 1990 Amendments based on the severity of ozone nonattainment. Those ratios vary between 1.1:1 for marginal ozone nonattainment areas to 1.5:1 for extreme areas for ozone precursors. Consequently, in an extreme ozone nonattainment area with a distance between the new source and emission reduction of 50 miles, a total offset ratio as high as 3:1 could be required.

IV. SOURCES OF ERCs

Emission reduction credits can be obtained in a variety of ways. These include process changes, emission controls, equipment shutdowns, and other more innovative techniques. As more offsets are sought and emission reductions used up, innovative techniques are more likely to be sought.

Process changes are becoming a more frequent source of emission reductions because of mandates under the Federal Pollution Prevention Act of 1990. This act requires industrial facilities to develop plans for reducing liquid, solid, and air pollutants from operations, and implementing those plans over the next 5 years. One example of emission reductions that have resulted from pollution prevention efforts is in changes that have been made from solvent degreasing operations to aqueous degreasers. Several new surfactants have been developed that can be used to remove oil, grease, or other contaminants from industrial parts. When a solvent degreaser is replaced with an aqueous degreaser, hydrocarbon emission reductions can be banked as emission reduction credits. Other pollution prevention process changes that can provide emission reduction credits include changes from solvent-based to water-based coatings and changes from internal combustion engines to electric motors in facility operations.

Emission controls beyond what is currently being used on processes or equipment can result in emission reduction credits. Usually the greatest opportunity for these offsets is provided with the oldest equipment around. Because of requirements under new source review that new emission units have BACT, reductions from additional controls, referred to as "over control," will be small. However, for equipment installed before 1977, few if any air pollution controls were required. Although retrofitting of controls is often difficult because of physical space limitations or the operating characteristics of the equipment, the difference in emissions between uncontrolled sources built before 1977 and BACT for new sources can be more than a factor of 10. Additionally, newer equipment is often more energy efficient so that the reduced emissions are also associated with lower operating costs.

A limitation on ERCs that can be obtained from control of older, uncontrolled sources is due to RACT requirements in nonattainment areas. Under these requirements, offset credits can only be obtained for emission reductions that are beyond RACT or other source-specific reductions already mandated in existing or proposed agency rules or in state implementation plans (SIPs) that have been approved. If a SIP provision affecting allowable reductions is anticipated, there may be benefit in installing controls on existing sources before such reductions are proposed so that ERCs can be maximized.

Even if a source has emission reductions mandated under RACT or SIP regulations, additional controls that might not otherwise be considered economical may be evaluated for obtaining emission reductions in a particularly tight offset market. Whether an "over control" emission reduction is economical is determined by comparing the price for overcontrol technology installation with the market price for ERC sales.

An obvious source of ERCs is facilities that are shut down. If a shutdown facility has emission units that have been operating during the past 2 years, the shutdown may provide opportunities for ERCs. ERCs that can be banked or sold are based on the actual emissions in the highest of the past 2 years, unless documentation can be provided that those emissions were not "normal." If such documentation is provided, the highest emission year in the past 5 years may be used as a basis for ERCs. The emission level for which the facility is permitted cannot be used as a basis for ERCs unless that was the actual level of emissions. Emissions greater than permitted levels cannot receive any ERCs. Arguments of less than normal emissions can be made based on economic or market constraints outside the control of the facility, or equipment failures that required extended time periods for repair.

A recent opportunity for banking ERCs has come from the shutdown of defense installations under Base Realignment and Closure (BRAC). When bases are closed, ERCs may be available to be banked for future use by entities taking over the closed property or for credits to offset increased defense activities elsewhere. One limitation on this second alternative is that the ERCs and emission increases must be in the same air basis for the credits to result in a "net air quality benefit."

If shutdowns are used for ERCs, the credits must be registered with the air pollution control agency within a limited time period, which can be as short as 90 days. One reason for this limited time period for banking ERCs is that it allows the regulatory agency to verify operating records for the facility while they are still available. Similar time limits on banking also apply to overcontrol emission reductions, but usually this is not a problem with overcontrol transactions because they rarely take place without prompt use of the credit created.

Offset transactions have been occurring for more than 15 years since the concept was introduced as part of SIPs in the late 1970s. Consequently, there are areas of the country where most conventional opportunities for emission reduction credits have been exhausted. In these areas, more innovative approaches have been pursued. One such innovation has been the purchase of old cars as an ERC. Several agencies have established rules for such transactions, including that the engine in the car be destroyed, documentation be provided of the number of miles driven by the car in the last year, and limitations on the duration of the ERC obtained of only 3 to 5 years. The logic of the last rule is that old cars would not run forever, and consequently an ERC for eliminating them should not either.

Another innovative approach to ERCs in areas with agricultural production is the use of agricultural burning credits. Agricultural burning has been traditionally used to get rid of waste prunings or other agricultural wastes. In recent years, cogeneration projects have been developed that utilize waste wood as fuel. By reducing the amount of open field burning, an emission reduction credit can be claimed. This credit can be considered permanent if contractual language is developed that assures a permanent prohibition (such as a deed restriction), but agricultural burning credits are limited because of seasonality. In agriculturally diverse areas, this limitation can be overcome by purchases of credits from producers of several different crops who burn waste on different schedules. Such transactions can become quite complicated.

A common source of offset credits for particulate matter emissions is paving of dirt roads or parking lots. ERCs cans be calculated for such paving projects, but the calculation requires information on the frequency of use of the road or parking lot that is being paved. Often data on use are not well documented, and consequently the offset cannot be demonstrated as being real. Nonetheless, this source of offsets is often popular in rural areas because of the social benefit of improved roads, and sketchy documentation of use may be accepted by the regulatory agency.

One of the provisions of the Clean Air Act Amendments of 1990 was that there be progressive annual reductions in ozone precursor emissions in ozone nonattainment areas. Air pollution control agencies have been left with the responsibility for implementing these reductions, and one likely opportunity for emission reductions will be older and uncontrolled sources. These same older and uncontrolled sources are good for securing ERCs. If older and uncontrolled sources are used as ERCs, the offset ratio applied will provide an opportunity for obtaining the required annual reductions in ozone precursors while also providing revenue to install controls on these sources. However, if the agency imposes an emission reduction requirement as part of a mandated rule, offset credit cannot be secured because the emission reductions are no longer surplus. Thus, with the implementation of the 1990 Amendments, opportunities for securing offset credits may increase as older and uncontrolled sources seek revenue to cover the cost of soon-to-be-required emission reductions.

V. OFFSET PRICES

Offset prices will be a function of several variables, only a few of which are at the control of a person seeking ERCs. The variables include the following:

- Supply and demand
- Previous transactions
- Control costs
- Location
- Terms of agreement
- Use of options

Offset prices are negotiated between a buyer and seller of ERCs in the context of supply and demand. Until recently, there have been too few ERC transactions to establish a market with multiple buyers and sellers. If there is only one buyer and multiple sellers of ERCs, the buyer can set the price. If there is only one seller, the seller sets the price, since whether there are multiple buyers or just a single buyer, the buyer is in need of the credits to go forward with a project, while the seller is already in business and needs the ERC transaction only to improve his or her income and lower his or her emissions.

An important determinant of ERC price is the price in previous transactions. The importance of previous transaction prices is the same as in a real estate market — the earlier price determines the approximate value of the product. Because previous prices are so important, both the price paid and the conditions of the sale need to be known. A law has been passed in California requiring disclosure of the prices paid for ERC transactions, and these data are published each year by each of the California air districts. In other locations, the price may not be "officially" disclosed, but there can be other reasons to disclose an ERC price, so the information is often available.

There is a relationship between the costs of emission reductions and the price of a transaction. Usually the lowest price that will be considered by a seller will be the cost of installing control technologies or process changes that will result in the emission reduction. As can be inferred from the discussion above about sources of ERCs, control costs can vary enormously. Generally, an emissions reduction technology will be more expensive to install if it is on an existing piece of equipment or process rather than part of a new installation. A particular concern is where there is inadequate physical space to install new equipment. Control cost estimating techniques are available,[2] or estimates can be obtained from vendors of the control equipment proposed. In some cases, a buyer may propose a lower purchase price than the cost of control equipment. This can occur when there are ancillary benefits to the pollution control equipment, such as energy savings. For example, a control technology for reducing NO_x emissions from stationary internal combustion engines (such as those used for natural gas compressor stations) not only reduces NO_x but increases the power output of the engine. Thus there is additional benefit to the seller.

Because offset ratios in may instances are a function of the distance and direction between a source of ERCs and the new source, the locations of ERCs will affect the price. Thus the value of a particular emission reduction will be different for buyers with new sources in different locations, and whether the new source is upwind or downwind of the emission reduction. If the new source is upwind, emission reductions may have no value at all, depending upon the regulatory agency rules.

The terms of an ERC purchase agreement also affect price. One proposal from a seller in a recent transaction specified that maintenance costs for the control equipment would be borne by the buyer for the life of the equipment. Such terms are often unacceptable to a buyer because so many maintenance costs for equipment that is part of another process can be due to causes beyond the control of the buyer, and thus could represent a substantial risk. In instances where such terms are demanded by the seller, prices are often lower, and insurance or a service agreement with the control system vendor can limit the uncertainty in financial exposure. Also, there is a potential problem for the seller if the buyer stops providing maintenance. The seller is constrained by his or her revised operating permit to emission limits based on installed control equipment. If a buyer does not maintain the equipment, the seller is subject to notices of violations and potential penalties.

Options can be used to adjust the timing for payment for ERC purchases, and to distribute the risk more satisfactorily to the buyer. Often options are proposed for the purchase of ERCs so that the source can be secured for a modest cost at a time when project financing is not yet complete. Often project financing is limited until the project has received all necessary permits and other approvals for construction. However, offsets must often be secured before final approval of a permit is granted. Thus there is a "Catch-22," which can be addressed by purchasing an option to buy the emission reductions at a later time. Usually this option has a time limit, after which the seller may seek other buyers, but during the period when the option is applicable, the buyer can demonstrate that needed emission reduction offsets have been secured. Options are generally priced at 5 to 10% of the purchase price, with a duration of 1 to 2 years, depending on the estimate of the approval time for the project (see Chapter 7 for a further discussion of schedule).

Prices for hydrocarbon ERCs have grown steadily from an average of $1,100 per ton per year in 1984 to over $13,000 per ton per year in 1995. The range of prices in a 6-year period in Los Angeles is shown in Figure 4-1. With this doubling of price in 18 months, ERCs would appear to be an asset that should be retained for future appreciation.

However, if the potential ERCs at a source that have not been banked become required as part of agency RACT requirements, they lose all value. Thus the decision on whether to sell ERCs becomes a risk decision weighing the appreciation of the value of the credits against the possible plummeting of their value to zero. As noted above, the number of buyers and sellers in an air basin containing

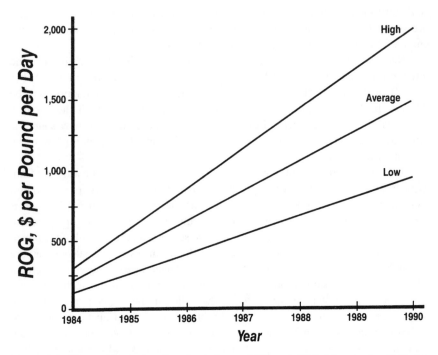

Figure 4-1 Offset prices in Los Angeles between 1984 and 1990. (Source: Aer-X Corpo-
        ration, 1990.)

older and uncontrolled sources will determine the market value of the emission
reductions.

Prices for NO_x ERCs range from about $4,000 per ton per year to as high
as $30,000 per ton per year. The highest values were for ERCs purchased in an
area with very few NO_x sources and consequently few opportunities for ERCs.

Recently, advancements in ozone chemistry models have indicated that
reductions in NO_x and VOCs are roughly equivalent in potential ozone concentration reductions. Consequently, agencies in ozone nonattainment areas have
been willing to accept interpollutant trades between VOC and NO_x ERCs at
approximately a 1:1 ratio for NSR permitting. This has equalized the price of
ERCs for these two pollutants in those areas.

As a result of the acid gas credits program established under Title IV of the
1990 Amendments, the market for sulfur dioxide emission reduction credits have
become quite stable. The Chicago Commodity Exchange has established a market
in acid gas ERCs and there have been numerous transactions in credits. The price
for SO_2 credits has remained at about $500 per ton since this market has been
established.

VI. CREATING ERCS

For an emission reduction to become an ERC, it must be approved by the
regulatory agency that is responsible for NSR. This approval will depend on

whether the emission reduction proposed satisfies the criteria of being real, quantifiable, surplus, permanent, and enforceable. Each of these has been discussed above, and the regulatory agency staff will need to be presented with documentation showing that the emission reduction meets each of these criteria.

Documentation of the reality and quantification of an emission reduction will include an acceptable estimate of the emissions before the reduction as well as proof that the reduction claimed can be achieved by the technology, process change, or other approach that is proposed. Use of well-documented emission factors, such as AP-42, can be proposed, but source testing results make a much stronger case for providing a source-specific emission factor. In addition to the emission factor, activity data are needed that document both the level of operation of the source and its hours of operation during the last 3 years.

Documenting that an emission reduction is surplus requires attention to current rules of the regulatory agencies overseeing the project as well as to air quality attainment plans or state implementation plan SIP for the geographical area in which the source is located. If a proposed reduction is mandated in current or proposed rules or the SIP, it cannot be used as an ERC.

Establishing the permanence and enforceability of an emission reduction can be straightforward for additional emission controls on a stationary source, but is much more difficult for the nonconventional emission reductions that are being proposed as stationary source reductions are used up. Deed restrictions may be used for agricultural burning offsets. Offsets that are only for a limited time period may be specified for reductions from old car purchases.

As emission reduction opportunities are eliminated by their use as offsets, creativity is needed by proponents for new sources. New technologies are continuing to be developed that provide opportunities for additional emission reductions, and for reductions in the emissions that are predicted from proposed sources. The combination of creativity and technology will continue to be important in obtaining ERCs.

REFERENCES

1. Irving, Patricia M., *Acidic Deposition: State of Science and Technology,* Vol. 1, Section 2 — Air Chemistry, B. B. Hicks, Principal Author, pp. 2–72.
2. Vatavuk, William M., *Estimating the Costs of Air Pollution Control,* Lewis Publishing, Boca Raton, FL, 300 pp., 1990.

5   AIR QUALITY IMPACT ANALYSIS

by Howard W. Balentine, CCM

I. INTRODUCTION

As noted in Chapter 1, an air quality impact analysis (AQIA) is required for PSD permitting and dispersion modeling is needed in order to conduct a health risk assessment. In this chapter, the use of dispersion models in air quality permitting will be presented. Dispersion modeling is an area that requires substantial training, and except for the simplest models available, professional expertise is needed. Therefore, the goal of this chapter will be to identify what is needed to conduct dispersion modeling, what information will be sought from the facility or project proponent, and how to assure quality in the modeling that is carried out.

The sections that follow will discuss fundamentals of dispersion modeling, the modeling protocol and how it is developed, input data required for modeling, selection of the most appropriate model, quality assurance, and sources of additional information.

II. OVERVIEW OF AIR POLLUTION METEOROLOGY

When air pollutants are released into the atmosphere, mixing occurs between the pollutants and the ambient air. Turbulence in the atmosphere governs the rise of plumes and the dispersion of pollutants in the plume into the ambient air. This turbulence is due to thermal and mechanical processes. Temperature differences between the surface and aloft result in vertical temperature stratifications that can induce convective buoyancy and turbulence during the day and inhibit convection and turbulence at night. Turbulence is also generated by the aerodynamic forces that result when the wind blows around and over obstacles and from the friction with the ground's surface. Under light wind speed, buoyancy effects in the atmosphere tend to dominate the production of the turbulence, whereas mechanical turbulence tends to dominate at higher wind speeds.

Atmosphere turbulence is not routinely measured directly because of the difficulty in doing so. Rather, estimates of the potential magnitude of turbulence are generally inferred from other meteorological measurements. The most commonly used surrogate for atmosphere turbulence is atmospheric stability class. Stability class A (or 1) is the most unstable condition and results in the greatest turbulence. Class A occurs only during the middle of the day under conditions of sunny skies with strong incoming sunlight and light winds. The most stable stability class, Class F (or 6), occurs only at night with light winds and generally clear skies so there is significant cooling to the night sky and the formation of temperature inversions. Midway between these two classes is neutral stability (Class D or 4). Stability Class 4 occurs with cloudy skies or with moderate to high wind speeds.

A number of techniques for estimating stability class from meteorological measures have been developed. One of the most common estimation schemes was developed by Turner.[1] In the Turner method, routine meteorological measurements of wind speed, wind direction, cloud amount, and cloud ceiling height are used along with the time of day to estimate stability class. In this scheme, the cloud cover values and the time of day are surrogates used to represent incoming solar heating during the day and nighttime cooling. Table 5-1 presents a summary of the Turner method for estimating stability class. Table 5-1 shows the dependence of unstable conditions on periods of low wind speeds and strong incoming daytime heating. At night, stable conditions are dependent on low wind speeds and clear skies. Higher wind speeds cause turbulence that prevents the formation of strong temperature contrasts between adjacent layers in atmosphere near the ground required for very unstable and very stable conditions. Consequently, higher wind speeds force the stability class toward neutral stability.

Table 5-1   Turner's Method for Estimating Stability Class

Surface wind speed (at 10 m), m sec^{-1}	Day Incoming solar radiation			Night Thinly overcast or	
	Strong	Moderate	Slight	≥4/8 Low Cloud	≤3/8 Cloud
<2	A	A-B	B		
2-3	A-B	B	C	E	F
3-5	B	B-C	C	D	E
5-6	C	C-D	D	D	D
>6	C	D	D	D	D

The neutral class, D, should be assumed for overcast conditions during day or night.

Source: Turner, D.B. 1970. "Workbook of Atmospheric Dispersion Estimates (Revised)". AP-26. Office of Air Programs, U.S. Department of Health, Education, and Welfare, Research Triangle Park, NC.

Under most normal dispersion conditions, vertical dispersion is confined to a specific layer of the atmosphere near the surface. This layer through which the vertical dispersion of pollutants occurs is called the mixed layer and the height of the top of the mixed layer is called the mixing height. The mixing height is a function of the time of day and season and can extend from near the ground to several thousand meters during periods of intense afternoon convection on hot and sunny summer afternoons. The mixing height is undefined during periods of stable meteorological conditions.

III. FUNDAMENTALS OF DISPERSION MODELING

The most commonly used model to simulate the transport and dispersion of a plume in the atmosphere is the "Gaussian" model. This model, derived from an analytic solution to the equation of motion in a fluid, is based upon the assumption that pollutants concentrations in a dispersing plume obey a Gaussian, or normal, distribution. In the Gaussian model, dispersion is a function of down-wind distance, with ever greater dispersion occurring as the plume is transported downwind on the mean wind flow. The coordinate system for pollutant dispersion and the Gaussian shape of the dispersing plume is shown in Figure 5-1. The fundamental Gaussian plume equation (Equation 5-1)[2] allows one to estimate the concentration of a pollutant at any point x downwind from the initial release, and at a distance z above ground level and y perpendicular to the centerline of the plume

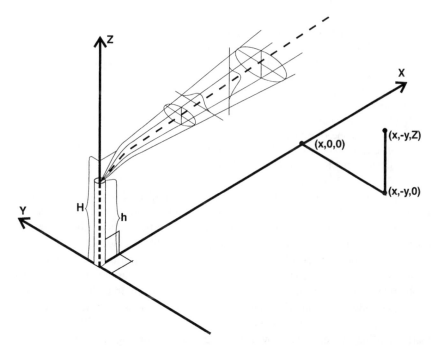

Figure 5-1 The coordinate system used for gaussian plume dispersion modeling. (Source: Turner, D.B. 1970. "Workbook of Atmospheric Dispersion Estimates (Revised)". AP-26. Office of Air Programs, U.S. Department of Health, Education, and Welfare, Research Triangle Park, NC.)

$$\chi(x,y,z:H) = \frac{Q}{2\pi\sigma_x\sigma_y u} \exp\left[-\frac{1}{2}\left(\frac{y}{\sigma_y}\right)^2\right]\left\{\exp\left[-\frac{1}{2}\left(\frac{z-H}{\sigma_z}\right)^2\right]+ \right.$$

$$\left. \exp\left[-\frac{1}{2}\left(\frac{z+H}{\sigma_z}\right)^2\right]\right\}$$

(5-1)

where χ = pollutant concentration, g/m^3, Q = emission rate of pollutant, g/sec, u = wind speed, m/sec, σ_y = dispersion parameter in the y-direction, meters, σ_z = dispersion parameter in the z-direction, meters, and H = effective stack height, meters.

Wind direction is not an input to Equation 5-1. The coordinate system established for the Gaussian model is a rotating coordinate system that has its X-axis always in the downwind direction. The calculation of the concentration using Equation 5-1 has been programmed into computer codes, called dispersion models, and so it is not necessary to go through it by hand.

In deriving Equation 5-1, a number of assumptions must be made that establish limits on the applicability of the Gaussian model. These assumptions include the following:[3,4]

- The meteorological flow is continuous, steady-state, and uniform, and the pollutant is released from a single source.
- The pollutant is released from a single elevated point source and the emission rate is continuous and constant.
- Turbulence is uniform in the horizontal and vertical directions, but varies with downwind distance.
- Downwind transport on the mean wind dominates downwind dispersion (i.e., the wind speed cannot approach zero).
- There is no loss of pollutants from the plume due to chemical reactions, deposition, or other physical processes.

Each of these assumptions is more or less realistic, depending upon the problem being addressed by the dispersion model. As a result of these assumptions, the Gaussian model is limited in applicability to those locations with relatively constant and uniform meteorological conditions. Because it if difficult to obtain uniform meteorological flow conditions in locations with significant terrain influences and channeling of wind flow, the Gaussian model is most applicable to areas of relatively uncomplicated, flat terrain. In addition, the model assumes steady-state conditions and is therefore not generally applicable to modeling instantaneous pollutant releases such as occurs during the spill of hazardous materials. The model does allow for changes in meteorological conditions such as turbulence or wind speed by superimposing concentration estimates made from successive applications of the equation using different sets of input conditions.

Typically, individual model concentration estimates are performed using meteorological and emission conditions representative of a 1-hr period. Because of the assumption of homogeneous, steady-state conditions, the range of applicability of the Gaussian model is commonly assumed to be 50 km.[5]

In the more sophisticated implementations of the Gaussian equation into a computer dispersion model, adjustments and modifications have been developed that overcome some of the limitations posed by the basic Gaussian model assumptions. Currently, models such as the EPA Industrial Source Complex[6] (ISC) model can accommodate multiple sources, complex terrain, nonuniform flow conditions, such as that which occurs in the aerodynamic wakes of buildings, and loss of pollutants through deposition.

Atmospheric stability class and downwind distance are not direct inputs to the Gaussian model. Rather, stability class and downwind distance are incorporated into the model through the dispersion parameters, σ_y and σ_z. These parameters are the assumed standard deviations of the horizontal and vertical plume Gaussian distribution. The most commonly used dispersion parameters for the Gaussian model are the Pasquill-Gifford-Turner (PGT) dispersion coefficients[7] for rural areas. Figures 5-2 and 5-3 present plots of the value of the PGT σ_y and σ_z coefficients as a function of stability class and distance downwind. For urban areas, the McElroy-Pooler coefficients[8] are commonly used. EPA has incorporated these two sets of dispersion coefficients into the regulatory versions of EPA Gaussian dispersion models.

Because a buoyant plume will rise to a certain height in the mixed layer, there will be some distance downwind before any of the pollutant can disperse to the ground. Generally, magnitude of the peak downwind concentration decreases and the downwind distance at which the peak downwind concentration occurs increases with increasing final plume height and increasing atmospheric stability (more stable conditions). The exception is for very unstable conditions, which because of the very large magnitude of turbulence, can bring essentially undiluted plume directly to the ground and produce peak concentration very near to the stack. This occurrence is called plume fumigation and occurs only during very specific combinations of meteorological conditions. The general decrease in downwind pollutant concentration with downwind distance is due to the dilution effect of mixing a given amount of pollutant plume into an ever increasing volume of ambient air.

For most regulatory applications, the magnitude of the peak downwind concentration and the location of the peak are the two primary results desired from a modeling analysis. The peak concentration is typically compared to an ambient standard or allowable PSD increment, while the location is used to judge if the peak concentration occurs in ambient air (e.g., outside the fenceline of the facility being modeled).

The Gaussian model is an imperfect representation of the chaotic nature of atmospheric transport and dispersion of pollutants. A "rule of thumb" typically ascribed to the Gaussian model is that it is accurate to within a factor of 2. There are a variety of reasons for this inherent limitation to the accuracy of Gaussian models, including uncertainty in input meteorological data, the difficulty in satisfying the assumption of a steady-state, homogeneous atmosphere, and uncertainty in source emission rate and plume release parameters.

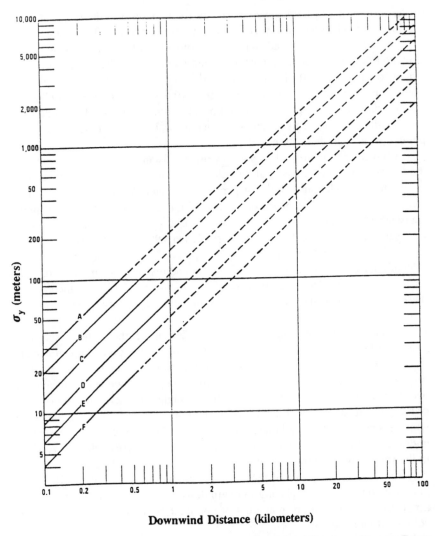

Figure 5-2 The dispersion parameter, σ_y, as a function of stability class. (Source: Turner, D.B. 1970. "Workbook of Atmospheric Dispersion Estimates (Revised)". AP-26. Office of Air Programs, U.S. Department of Health, Education, and Welfare, Research Triangle Park, NC.)

In the application of dispersion models based on the Gaussian approximation, conservative assumptions are generally made. By conservative, it is meant that, given the choice, assumptions are made that will tend to cause modeled concentrations to be overpredicted. However, for a given modeling situation involving a single source and a specific set of meteorological conditions, the model can as readily underpredict as overpredict the mean concentration of the pollutant. Therefore, it is not correct to assume that, since EPA regulatory models are "conservative," they will always overpredict ambient pollutant concentrations.

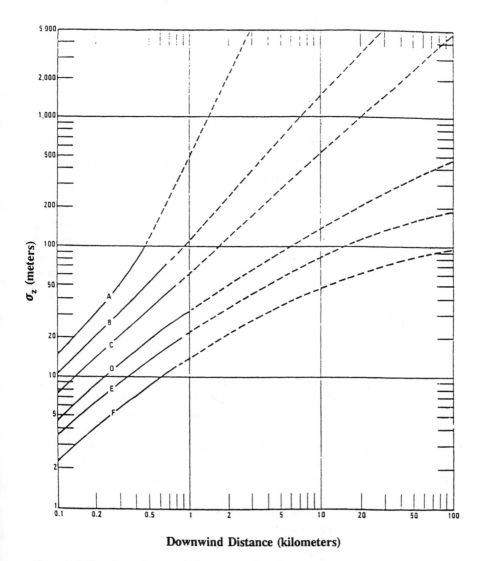

Figure 5-3 The dispersion parameter, σ_z, as a function of stability class. (Source: Turner, D.B. 1970. "Workbook of Atmospheric Dispersion Estimates (Revised)". AP-26. Office of Air Programs, U.S. Department of Health, Education, and Welfare, Research Triangle Park, NC.)

IV. MODELING PROTOCOL

The development of a modeling protocol is the first step in a dispersion modeling analysis performed as part of a regulatory permitting analysis. The modeling protocol describes how the modeling will be carried out. It is usually necessary to submit a modeling protocol to the regulatory agency in order to receive agency approval for the modeling prior to performance of the modeling. The modeling protocol contains the following elements:

- Dispersion model selection
- Modeling methodology and assumptions
- Source input selection and preparation
- Meteorological data selection and processing
- Basis for ambient pollutant concentrations
- Receptor grid development

A. Dispersion Model Selection

AQIAs are carried out using Gaussian dispersion models — models that use the basic assumption of Gaussian dispersion described in Section III above. There are several such models, and different models are applicable to different source types and terrain.

The primary reference for selection of models is EPA's *Guideline on Air Quality Models*.[9] It identifies the models that have been developed and validated by EPA and are considered reliable for use in permitting. In particular, there are two models included in the *Guideline* that are most commonly used. They are:

- SCREEN3
- ISC3

1. *SCREEN3*

SCREEN3 is the latest (1995) version of an EPA model used to screen sources for potential worst-case concentrations.[10] SCREEN3 is a very easy-to-use model that can be applied to a single source and provides estimates of the worst-case (highest) impacts expected from that source's emissions. If the generally conservative results obtained with SCREEN3 are able to demonstrate that impacts from a proposed project are insignificant, no additional modeling is typically needed. SCREEN3 allows estimates of pollutant concentrations to be made for a single point or area source in areas of flat to complex terrain. In dispersion modeling terms, complex terrain is defined as terrain whose elevation exceeds the release height of the sources being modeled.

2. *ISC3*

Industrial Source Complex Version 3 (ISC3) is the most frequently used EPA regulatory dispersion model. Along with SCREEN3, it was released in 1995 and replaces the previous version (ISC2) as the primary regulatory model. ISC3 is a refined model, in EPA terminology, in that it allows detailed assessment of pollutant concentrations from multiple sources and for different averaging times in flat and complex terrain areas. If used with the worst-case meteorological data from SCREEN3, ISC3 gives the same results for a single source.

ISC3 is modular in design and was designed by EPA for ease in maintenance and modification as new modeling techniques become available. The primary enhancements over ISC2 (the previous version) incorporated into ISC3 are refined treatment of area sources, the capability to model sources in complex terrain, and new treatment of deposition.

While other models are identified in the *Guideline,* the combination of SCREEN3 and ISC3 should be satisfactory for most regulatory modeling situations involving typical point and area sources and for modeling distances out to 50 kilometers. With the incorporation of complex terrain modeling capability into ISC3, older *Guideline* models for complex terrain such as the COMPLEX-I model and Complex Terrain Dispersion Model Plus (CTDM-Plus) no longer need to be used in conjunction with ISC to estimate pollutant impacts in complex terrain areas. Consequently, there should be little difficulty encountered in justifying to the regulatory agency the use of the ISC3 model for a typical permitting analysis.

Beyond 50 km downwind of a source, the ISC3 model is not considered to be a reliable estimator of ambient pollutant concentrations. Other modeling techniques are required to estimate pollutant concentrations at these far downwind distances. The principal reason for estimating pollutant concentrations beyond 50 km is to assess potential impacts of the source in sensitive areas such as National Parks and other Class I areas under the Prevention of Significant Deterioriation program. For these situations, there can be the need to assess potential visibility impairment, ambient pollutant concentration, and pollutant deposition. Guidance is given in the *Guideline* and other EPA documents on the appropriate models and modeling techniques to be used in assessing pollutant impacts in long-range transport situations. In general, because each long-range transport impact assessment includes site-specific issues, the model selection and application methodology must be negotiated with the appropriate agency as part of the modeling protocol development process.

EPA, in conjunction with other federal governmental agencies and the American Meteorological Society, has embarked on a model development program with a goal of enhancing, and eventually replacing, the ISC model.[11] The latest version (ISC3) is the second major revision to the model in the past 3 years. All aspects of the model are under review, including replacement of dispersion coefficients with direct measures of turbulence, updated estimates of plume rise and plume penetration through the mixing height, and better treatment of diffusion in convective and stable layers near the ground.[12]

3. Model Availability

Computer codes for all of the guideline models are available on EPA's Support Center for Regulatory Air Models (SCRAM) Bulletin Board,[13] the National Technical Information Service (NTIS), and from several consultants who have added features to models they sell to make them easier to use. Several of these consultants also provide training in model use.

The EPA SCRAM Bulletin Board is maintained by EPA's Office of Air Quality Planning and Standards (OAQPS). It provides not only dispersion models, but also modeling guidance and meteorological data for different geographical areas. It is one of a number of EPA bulletin boards that are part of EPA's Technology Transfer Network (TTN).

B. Modeling Methodology

A number of options are available to the modeler in the selection of modeling methodology when applying a model such as ISC3. These options include such things as specific model options selected, quality assurance, and results presentation and reporting. The modeling protocol should address each of these areas in sufficient detail to allow the regulatory agency to determine the acceptability of the modeling approach.

A variety of "switches" select processing options in the ISC3 model. The EPA has defined a specific set of these switches as regulatory default values. If any change is made to these regulatory default switches, detailed justification should be presented in the modeling protocol and negotiated with the regulatory agency.

A key component of a modeling analysis is the performance of quality assurance on the input and output data to ensure the integrity of the modeling results. The amount and type of quality assurance performed will depend upon the specific modeling application and the regulatory agency. Typically, good quality assurance procedures include verification of all input data by a second modeler, preparation of graphical and tabular data to compare model input and output with expectations, verification of model input data against basic sources of information, and general review of all model results for reasonableness.

The format to be used to present the model results should be defined. Typically, in addition to a written report, the agency will require submittal of hard copy and electronic copies of model input and output files. The specific components of the report should be negotiated with the agency during the protocol approval process.

C. Source Inputs

Source inputs include:

- Emission rate
- Source location and elevation
- Stack height
- Stack diameter, gas exit speed and temperature (for point sources) or source dimensions (for area sources)
- Dimensions of adjacent buildings

1. Emission Rate

Emission estimating techniques were described in Chapter 2. For the modeling, it is necessary to make emission estimates for each of the averaging times used in the modeling. Annual average emission estimates can take into account scheduled outages of a plant, such as for annual maintenance, as well as the scheduled daily or weekly operation anticipated. A number of hours of operation per year needs to be selected with care because this assumption is likely to become a permit condition. If there are no consequences of a high annual emission estimate, 8,760 hours of operation can be assumed. However, this assumption will result in a higher estimate than an estimate that might represent the maximum practical hours of operation. Possible consequences that need to be considered include fees that might be based upon permitted limits or annual ground level pollutant concentrations that exceed an annual average air quality standard.

Shorter averaging times may have higher emission levels than those used for annual emissions because of short-term operations that have emission rates that cannot be sustained for an annual period. An example of this might be batch operations.

The emission levels selected for modeling are important because these are the levels that will be used to determine the air quality impacts of the proposed project. Consequently, these are the levels that will become conditions in the resulting permit.

2. Source Location

The location of the source is important because it will be used to determine the surrounding terrain and land use, the relative location of other emission sources not part of the proposed project, and the distance to the nearest PSD Class I area. The location will also establish the attainment status of the area, but this information will probably have already been determined as part of the regulatory analysis described in Chapter 1. USGS 7 $^1/_2$ minute quadrangle maps are often required to be submitted with a permit application, and they are helpful in determining terrain in the vicinity of a proposed project on a scale comparable to the scale of dispersion modeling.

3. Stack Height and Diameter

A pollutant plume released at an elevated temperature will tend to rise in the atmosphere due to buoyancy and mix with ambient air. Likewise, if the plume has initial upward momentum, it will rise upward under the influence of momemtum. Eventually, the plume reaches a height at which the upward momentum of the plume has been dissipated due to turbulence and the temperature of the plume has become the same as the ambient air due to mixing. When this occurs, the plume stops rising and the plume is said to be at its final plume rise height. The sum of the initial release height of the plume and the amount of plume rise is called the effective stack height.

As indicated in Figure 5-1 and Equation 5-1, stack height is an important parameter in determining both the dispersion that occurs and the location of the point of maximum impacts. Often there is the possibility of changing the stack height for a proposed project, and increasing stack height can reduce maximum impacts. Stack height may also have to be decreased to meet restrictions from the Federal Aviation Administration if the proposed stack is near an airport.

Although increasing stack height reduces pollutant concentrations at the point of maximum impact, there is a limitation to the maximum stack height that can be used called the Good Engineering Practice (GEP) stack height. The concept of GEP was included in regulations after several facilities were proposed with extremely high stacks (on the order of 1000 feet). These stacks were not considered the best way to reduce the impacts of air pollution, since the pollutant emissions were just as large, but were dispersed over a wider area. The GEP stack height is defined as 2.5 times the height of any adjacent building or structure. A higher stack can be constructed, but for regulatory modeling purposes, only the GEP height may be input to the dispersion model. A stack shorter than GEP must be modeled at its actual height.

4. Stack Diameter, Gas Exit Speed, and Temperature

For point sources, stack diameter and exit speed are determined by the flow rate of flue gas. For area sources, stack exit parameters are not applicable. For area sources, the needed input parameters are the two side dimensions of a rectangular area representing the area source.

Using a larger induced draft (ID) fan can increase the plume exit speed for a given stack diameter for a point source. The selection of diameter and exit speed then must be part of the design of the process. A higher exit speed will increase the rise of the plume after it leaves the stack, and thus improve dispersion. The cost of the larger ID fan to accomplish this will need to be weighed against the possible benefits of lower maximum ground level concentration at the point of maximum impact.

Stack gas temperature is also a design parameter for the process. A higher exit temperature will increase the effective stack height because of thermal rise of the plume. The cost of this benefit is lost thermal energy which may have been usable heat in the process.

Rain caps also must be considered in stack design and subsequent modeling. Various types of rain caps are shown in Figure 5-4. From the point of view of pollutant dispersion, a concentric cylinder rain cap is definitely preferred. The logic of the operation of the concentric cylinder is that rain almost never falls straight down. It falls at some angle, and consequently will hit the rain cap and run down the outside of the stack. For small sources, a horizontal rain cap is more common. While a horizontal cap will not entirely eliminate vertical momentum from the plume and does nothing to the plume buoyancy, regulatory agencies many times require the assumption of zero plume rise when a horizontal cap is in place on the stack. When a horizontal rain cap is used, the only way to increase effective stack height for modeling purposes is to increase stack gas temperature, and hence the resultant buoyancy of the plume (if allowed by the regulatory agency).

5. Dimensions of Adjacent Buildings

The dimensions of adjacent buildings are critical in the dispersion of pollutants from a stack. Aerodynamic wakes in the lee of buildings and other structures can greatly increase pollutant concentrations near a source. In order to define the extent of potential building wakes that affect a given source, the length, width, and height of all structures adjacent to the source must be determined. The EPA

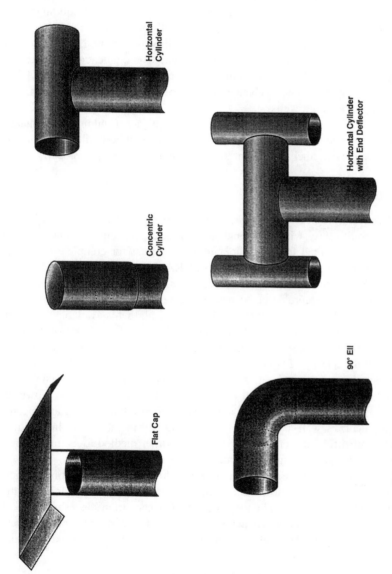

Figure 5-4 Different types of rain caps used for stacks. (Source: Radian International LLC.)

has developed the Building Profile Input Program[14] (BPIP) to automate the process of computing the required model input building wake parameters for modeling. Using the BPIP program, wake parameters are computed for each of 36 wind directions around each source.

Unfortunately, there are several kinds of nonstandard buildings and structures that are often present at an industrial site. These include tiered buildings, pipe racks, scaffolding, and tank farms. To successfully model an actual site, a site visit is helpful for the modeler, and if the facility is proposed, plot plans are important to use as a basis for modeling input. Communication between the permitting team and the design team is important for this part of the modeling protocol development, since a change in layout of the plant can have a substantial effect upon modeling results.

Because of the sensitivity of modeling results to adjacent buildings, information on what the site layout will be and how building wake effects will be included in the modeling is extremely important to include in the modeling protocol.

D. Meteorological Data

The meteorological data required for dispersion modeling includes mixing heights, stability class, wind speed and direction, and temperature. Meteorological data must be provided for every hour during the year for complete modeling. These data can be obtained from the EPA SCRAM Bulletin Board, the National Climatic Data Center, local regulatory agencies, or from on-site monitoring. In many instances, meteorological data may be fairly old. It is not unusual for meteorological data from the 1960s or 1970s to be all that is available. Although these data may be acceptable to the regulators, for a particularly sensitive application, there may be benefits to collecting current data because of public skepticism about the quality of "old" scientific data.

On-site meteorological data are the most reliable for modeling, and usually predict lower concentrations than use of artificial (worst-case) data. However, collecting these data requires a capital investment for both measuring and recording equipment, as well as the cost of operation and maintenance. In addition, EPA requires different methodology from the Turner method to estimate atmospheric stability class using on-site meteorological data.[15]

While on-site monitoring has the potential to produce lower estimated pollutant concentrations than artificial or remote data, the delay of a year before the local data set is complete can be very costly to a project. In instances where local data would help, a regulatory agency will sometimes be willing to consider local results from one quarter or half the year to provide an indication of likely annual results.

Different calculation schemes have been used, but the most common method is to assume that the other three quarters will be the same as the first quarter, or to use the first quarter along with artificial data for the remainder of the year. Often, regulatory approval can take nearly a year, so prompt installation of a meteorological monitoring station can result in collection of a quarter or a half year's data before the dispersion modeling must be performed. Typically, if less than a year's data are used in the modeling, the agency will require performance and submittal of the results of a revised analysis once the complete year of monitoring data are available.

E. Basis for Ambient Pollutant Concentrations

There is usually not much choice in the selection of ambient pollutant concentration data used to represent background pollutant concentrations in the modeling area. Monitoring is conducted in each jurisdiction to determine which pollutants are attainment and which, if any, are nonattainment. However, sometimes there are very few monitoring stations in an area, and particularly in complex terrain areas there may be reason to believe that concentrations in the vicinity of the proposed facility are not the same as in other locations. Convincing a regulatory agency that a project in a nonattainment area should be subject to rules for attainment areas because of local monitoring results is unusual, but there is a better possibility for using local ambient concentrations as the baseline to which predicted increases from the facility are added. If local data indicate much cleaner air than other measuring stations in an area, this can provide an additional buffer of allowable increases from the proposed project. The disadvantage of this approach is that ambient monitoring station installation and operation costs are higher than those for a meteorological station.

F. Receptor Grid Selection

As part of the modeling protocol, it is necessary to identify the receptor points for which modeling results will be calculated. The regulatory agency will want to be sure that you have selected receptors that are likely to include the point of maximum impacts from a proposed project, and the applicant wants to minimize the number of points so that modeling costs will be minimized. Modeling is usually done in at least two steps. There are four groups of receptors for which concentrations are computed. These are:

- Coarse grid receptors
- Fine grid receptors
- Fenceline receptors
- Discrete receptors

When modeling in other than flat terrain areas, receptor elevations are required for all receptors. These elevations can be obtained from USGS 7.5 min maps of the modeling area, electronically from USGS digital data sets, or other sources. When using digital terrain data, care must be taken in gaining approval from the regulatory agency. Digital data, because they are typically derived from data at a coarser resolution than 7.5 min, are smoothed and display less variation than those obtained by hand from a 7.5 min quadrangle map. Consequently, terrain peaks will be lower when using digital data. Without prior approval, these differences may cause problems if the agency compares the stated elevations with those pulled by hand from a 7.5 min map as a quality assurance step.

1. Coarse Grid Receptors

The receptors are first set out in a coarse grid surrounding the facility. Such a coarse grid is shown in Figure 5-5. The purpose of the coarse grid is to identify the general locations of large ground-level concentrations. Typically coarse grid receptors are spaced at 200 to 1000 meter intervals, or greater.

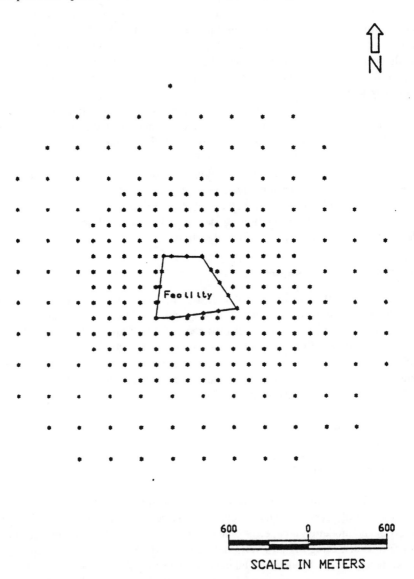

Figure 5-5 Receptor grid spacing for modeling around a facility. (Source: Radian International LLC.)

2. Fine Grid Receptors

The second step uses a fine grid that looks in more detail at areas where concentrations are high. Typically fine grids are at 50 to 100 meter spacing, but in some cases, lower spacing may be requested by the regulator. The purpose of the fine grid is to identify the maximum ground-level concentration point, and to identify local gradients in concentrations. In the modeling protocol, a commitment is made to conduct fine grid receptor modeling, but since the locations of high concentrations on the coarse grid are not yet known, the locations for specific fine grid cannot be proposed in the protocol.

3. Fenceline Receptors

Additional receptors are located at the fenceline for the facility to address possible building downwash effects. Inside the fenceline, ambient concentrations are not evaluated since the area within the fenceline is not accessible to the public, and is therefore not ambient air. Thus, if there are high concentrations due to building downwash, they would only be important if the concentrations at or beyond the fenceline are high.

Fenceline receptors are typically spaced 25 to 100 meters apart, depending on the modeling situation, the proximity of emission sources, and the requirements of the regulatory agency.

4. Discrete Receptors

In addition to finding the point of maximum concentration and obtaining the information to plot isopleths of concentration, the regulatory agency may require computation of concentrations at discrete receptors such as schools, day care centers, hospitals, and nursing homes. These are locations of sensitive populations, which may be more susceptible to inhalation exposure. This is discussed further in Chapter 6 on hazardous air pollutant permitting requirements.

G. Protocol Submission

Before the protocol is submitted, it can be worthwhile to make both coarse and preliminary fine grid modeling runs to see if concentrations predicted are likely to exceed the regulatory limit — ambient standard or PSD increment. There may be opportunities to specify a different fenceline boundary if nearby concentrations are too high, or it may be possible to rearrange sources within a proposed facility so that sources with high emissions are more distant from sensitive receptors or the facility fenceline. These adjustments can be made after the protocol has been approved, but will require more interaction with the regulatory agency than some applicants are comfortable with.

The protocol is reviewed by the regulatory agency modeling staff to be sure it meets the agency criteria for completeness, use of approved models, and other requirements. Typically review of a modeling protocol is completed within a few weeks — there are no regulatory limits set on protocol review time by most agencies. Interaction between the applicant's modeler and the agency modeler are recommended, including discussion prior to protocol submittal. Often uncertainties or apparent errors that appear to the agency modeler in reading the protocol can be clarified or corrected in conversation between professionals in this field. In addition, the agency modeler may give an indication of the intent to approve the modeling protocol some time prior to receipt of that approval in writing. If this happens, the modeling can begin after oral approval has been indicated so that the application can be completed more promptly.

V. MODELING RESULTS

When modeling results are available, there are several considerations which are important in interpreting them. These include:

- Uncertainty in model estimates
- Time and space comparisons
- Number of significant figures
- Averaging time conversions
- Quality assurance

A. Uncertainty in Model Estimates

Air quality dispersion models are used to forecast whether a proposed facility will cause air pollutant concentrations to exceed ambient standards or allowable increments prescribed under PSD. Because this is their use, the regulatory guidelines for application of the model tend to include inherent conservatism. The regulatory-specified modeling methodology is generally designed to overpredict concentrations (be conservative) on an average basis. However, for any given source–receptor modeling pair for a given set of meteorological conditions, the model can as easily underpredict as overpredict the true (expected) concentration. Therefore, one must be careful in assuming the ISC3 model will always produce conservative results, particularly when examining the results for a single source–receptor pair. The conservatism in model predictions is accepted because the purpose of the regulatory process is to protect public health, and there must be a guard to protect the public from uncertainty in model results.

B. Time and Space Comparisons

Gaussian dispersion models are most accurate in predicting the value of a maximum concentration. They are less accurate in predicting the physical location of that maximum concentration and when that maximum may occur during the operation of the facility. The models calculate concentrations based on historical records of wind and temperature patterns in the vicinity, not on actual winds and temperatures during the operation of the plant. Thus, although the model can produce reasonable estimates of the maximum concentration that can be expected to occur, the actual location of the occurrence of that maximum concentration or the time during the year when it will occur are very uncertain. This can be a problem in the calculation of health risk from a proposed facility and the tendency of the public to interpret spatial health risk results with respect to their own home or their child's school. This will be discussed further in Chapter 6.

C. Significant Figures

Dispersion modeling results are considered valid, within the constraints discussed above, to two significant figures. This can be misleading, since the computer that carries out dispersion calculations will present results commonly to eight digits. However, this is only an indication of the quality of the calculator, not of the quality of the model or inputs to that model. In accepting two significant figures, there has nonetheless been disagreement about how rounding should be made to get the second figure. Those seeking additional conservatism in the results argue that rounding up should always be done, while statistical rigor argues for rounding down if the third digit is 1 through 4, and up if the third digit is 5 or above.

D. Averaging Time Conversion Factors

A simplifying technique that has developed in the modeling community has been to run models to obtain concentrations for a 1-hr averaging time, and then use conversions to obtain the results for 24-hr or annual results. The conversions that are accepted are shown in Table 5-2. However, with ISC3, it is a simple matter to directly estimate concentrations for all averaging periods of interest from 1 hr to the annual period from a file of hourly concentrations at each receptor for the full year.

Table 5-2   Conversions of Dispersion Results to Different Averaging Times

Calculated impact	Conversion factor	Averaging time
1-hr average	1.0	1-hr average
	0.7	8-hr average
	0.4	24-hr average
	0.1	Annual average

E. Quality Assurance

Quality assurance in modeling is as important as in any other elements that are prepared as part of a permit application. Because the modeling effort is a computer computation, checking results must be approached differently than with a hand calculation, but it is still an important part of the process of completing the application. Elements of quality assurance discussed here include:

- Input review
- Emission estimate checking
- Graphical checks of source locations
- Concentration isopleths of results
- Review of "worst case" meteorology
- "Laugh test"

1. Input Review

The output of a guideline model includes a printout of the input parameters that were used in the computations. These parameters should be checked against the inputs provided to the modelers. Often, it is necessary to change units between information given to the modeler and the inputs necessary for the model, since the model requires inputs in metric units. Two common errors in modeling result interpretation are to assume that the actual emission rate is used in the modeling and that the results apply for the averaging time of interest. The first error occurs because modeling is often accomplished assuming a 1 g/sec emission rate. The modeling output is then in $\mu gm/m^3/g/sec$, sort of an emission factor for ambient concentration determination. This factor must be multiplied by the emission rate (in g/sec) to obtain the resulting ground-level concentration. The second error is to assume the concentration computed is for the averaging time of interest — in short, to ignore Table 5-2. It is a bit of good fortune that concentrations for longer averaging times are lower than the 1-hr averaging time result given by the model, so the answers are actually lower than the value seen in a superficial inspection. However, considerable anxiety can be avoided by remembering to consider the averaging time is 1 hr in some raw model output.

Another common error is to assume that because a modeling result is favorable — shows that the proposed project does not violate any ambient standards or PSD increments — it must be right. It is much easier to generate the zeal for close checking of a modeling output if you do not like the results than if you do.

2. Emission Estimate Checking

Although a considerable effort usually goes into calculating emission estimates for a project, as discussed at length in Chapter 2, errors are still possible. There should be a clear trail from a source test result or other basis to the emission estimate for the proposed project. Even if the estimating technique relies on inferences from similar processes or other such estimates, the assumptions made and supporting calculations must be traceable from start to finish. In this manner, assumptions are clearly stated, and consideration of the uncertainty in the estimates will be much more apparent. This is especially a problem when two different organizations have been involved — perhaps one estimating the emissions and another doing the modeling.

3. Graphical Checks of Source Locations

All of the inputs to the computer are digital, including the coordinates of the locations of emission sources. A good way to check the accuracy of input of these data is to plot on a graph the locations of each source. The computer can actually be asked to do this, using the input data to locate sources. This can then be compared with a plot plan to verify that each source is correctly located.

4. Concentration Isopleths of Results

Another means of graphically checking the results of modeling is to generate concentration isopleths predicted by the model. Really bizarre shapes to concentration isopleths should be associated with unusual topography or other characteristics of the facility or its surroundings. If such an association cannot be made, the modeling should be reviewed in more detail to determine whether there have been errors.

5. Review of "Worst Case" Meteorology

Another graphic check that should be made using the isopleths described in the previous paragraph is to be sure that the highest concentrations are downwind of the source. Since the wind direction is not constant all year long, the highest annual concentration could logically be in any direction. However, there is almost always a prevalent direction of wind for an area, and the highest concentration would be expected to be in that downwind direction. Another check is to substitute a "worst case" meteorological data set for the actual meteorological data used to see what effect this has on the maximum concentration. If the actual meteorological data calculates a higher concentration than the "worst case" meteorological data, there is probably an error.

6. Laugh Test

Finally, if the results of the modeling predict concentrations that are impossible to believe, they are probably wrong. Although modeling results can be difficult to interpret because they are nonlinear with respect to meteorological inputs, they are linear with emission rate. Thus a degree of "seat-of-the-pants" engineering is still appropriate, and modeling results should not cause laughter by reviewing agency staff.

REFERENCES

1. Turner, D.B. 1964. "A diffusion model for an urban area," *J. Appl. Meteor.*, 3(1) 83–91.
2. Turner, D.B. 1970. "Workbook of Atmospheric Dispersion Estimates (Revised)". AP-26. Office of Air Programs, U.S. Department of Health, Education, and Welfare, Research Triangle Park, NC.

3. Viegele, W.J., and J.H. Head, 1978. "Derivation of the Gaussian Plume Model," *J Air Pollution Control Association,* 28(11), 1139–1141.

4. Wilson, R.B. 1993. "Review of Development and Application of CRSTER and MPTER Models," *Atmospheric Environment* 27B(1), 41–57.

5. U.S. Environmental Protection Agency, 1995a. "Guideline on Air Quality Models (Revised)". EPA-450/2-78-027R, Office of Air Quality Planning and Standards, Research Triangle Park, NC.

6. U.S. Environmental Protection Agency, 1995b. "User's Guide for the Industrial Source Complex (ISC3) Dispersion Models (Revised)". EPA-454/B-95-003a, Office of Air Quality Planning and Standards, Research Triangle Park, NC.

7. Turner, D.B. 1970.

8. Gifford, F.A. 1976. "Turbulent Diffusion-Typing Schemes: A Review," *Nuclear Safety* 17(1), 68-86.

9. U.S. Environmental Protection Agency, 1995a.

10. U.S. EPA, 1995c. "SCREEN3 Model User's Guide," EPA/4-92-006, Office of Air Quality Planning and Standards, Research Triangle Park, NC.

11. Weil, J.C. 1992. "Updating the ISC Model through AERMIC," Paper 92-100.11, 85th Annual Meeting of the Air and Waste Management Association, Kansas City, MO.

12. Perry, S.G., A.J. Cimorelli, R.F. Lee, R.J. Paine, A. Venkatram, J.C. Weil, and R.B. Wilson, 1994. "AERMOD: A Dispersion Model for Industrial Source Applications," Paper 94-TA-23.04, 87th Annual Meeting of the Air and Waste Management Association, Cincinnati, OH.

13. The SCRAM computer bulletin board is one of several bulletin boards operated by the Office of Air Quality Planning and Standards as part of the Technology Transfer Network (TTN). The objective of the TTN is to promote the exchange of information between U.S. EPA, the States, and the public. The telephone number for the bulletin board is 919-541-1447. A voice help line can be reached during normal business hours in Research Triangle Park at 919-541-5384.

14. U.S Environmental Protection Agency, 1993. "User's Guide to the Building Profile Input Program," Office of Air Quality Planning and Standards, Research Triangle Park, NC.

15. U.S. Environmental Protection Agency, 1995a.

6   TOXIC AIR POLLUTANT ANALYSIS

I. INTRODUCTION

Toxic air pollutants are trace chemical compounds present in the air that are harmful to humans. Concern for toxic air pollutants was first expressed in the Clean Air Act of 1977. In that act, Section 112 defined a process for identifying hazardous air pollutants (HAP) and established a structure for determining the national emission standards for hazardous air pollutants (NESHAP). Although most regulated air pollutants can be considered toxic air pollutants because they are harmful to humans, toxic air pollutants are usually limited to non-criteria pollutants that cause cancer or have other acute or chronic health effects. The greatest focus in the area of toxic air pollutants has been on the cancer-causing pollutants, although various statutes have broadened coverage to apply to acute or chronic noncancer-causing pollutants as well.

Toxic air pollutants are also distinguished because their effects extend over limited areas around sources of each pollutant. Strictly speaking this is true for benzene, which causes cancer in laboratory animals, but benzene is emitted in trace quantities from most combustion processes, including automobile engines. Consequently, concentrations of benzene can be measured in the air in most urban areas.

The other distinguishing characteristic of toxic air pollutants is that both emissions and ambient concentrations are low compared to the emissions and concentrations of criteria pollutants. Typically toxic air pollutant emissions are measured in pounds rather than tons per year, and the health effects which can result are caused by very small concentrations.

The first step in determining what requirements must be satisfied in air quality permitting regarding toxic air pollutants is to determine what toxic air pollutants may be emitted from a process, and to estimate those emissions. This is the first topic discussed in this chapter.

The next step is to evaluate toxic air pollutant controls or impacts. The two approaches that are being used for this evaluation are based on technology standards and risk standards. Each of these is discussed below.

II. TOXIC AIR POLLUTANT EMISSIONS AND REGULATIONS

A. Toxic Air Pollutant Emissions Estimates

Toxic air pollutant emissions are estimated using the same techniques described in Chapter 2. There are several compilations of air pollutant emission factors that both identify and provide ways of estimating toxic air pollutant emissions from a particular process.

There are larger uncertainties in estimating toxic air pollutant emissions than in estimating criteria pollutant emissions. Since toxic air pollutants are either trace metals and other compounds that come from fuels or feedstocks entering a process or are products created during a process, such as products of incomplete combustion (PICs), there is less consistency in emissions of toxic air pollutants than of criteria pollutants when the same process uses fuels or feedstocks from different sources.

Emission estimation uncertainties also stem from the small number of tests that have been conducted to measure toxic air pollutant emissions. The testing conducted under the requirements of the California "Air Toxics 'Hot Spots' Information and Assessment Act of 1987" (AB 2588) increased by an order of magnitude the source test results available on toxic air pollutant emissions from stationary sources. These data are becoming available in AP-42 as well as other sources.

If the control equipment or other conditions imposed on a new or modified source based on literature estimates of toxic air pollutant emissions will be expensive to implement, there may be value in conducting source tests on the new or modified source or its equivalent. By definition, a new source permit is for a source not yet in operation, but the design for this new source is often based on existing equipment somewhere. Source tests using that equipment with the feedstocks and fuel sources planned for the proposed source will provide more representative data than the literature and may show that emissions are substantially different than estimates from literature emission factors. In some jurisdictions, research testing can be conducted if the applicant is willing to incur the expense of installing necessary, usually temporary, equipment and performing testing. Of course, there is a risk that the measured emissions can be substantially higher than literature estimates, and even more onerous controls are required, but knowing this in advance of construction rather than discovering it when compliance testing is carried out will probably result in lower costs to the applicant.

B. Toxic Air Pollutant Regulations

Title III of the 1990 Clean Air Act Amendments increased the number of HAP from eight, which had been identified in earlier rulemaking, to 189, including 168 individual compounds and 21 compound categories. One pollutant has since been dropped from the list. The list of HAP is shown in Table 6-1.

Major HAP sources are stationary sources or groups of sources within a contiguous area and under common control that have the potential to emit more than 10 tons per year of any single HAP or 25 tons per year of any combination

Table 6-1 Section 112 Hazardous Air Pollutants

Chemical abstracts service number	Pollutant
75-07-0	Acetaldehyde
60-35-5	Acetamide
75-05-8	Acetonitrile
98-86-2	Acetophenone
53-96-3	2-Acetylaminofluorene
107-02-8	Acrolein
79-06-1	Acrylamide
79-10-7	Acrylic acid
107-13-1	Acrylonitrile
107-05-1	Allyl chloride
92-67-1	4-Aminobiphenyl
62-53-3	Aniline
90-04-0	o-Anisidine
1332-21-4	Asbestos
71-43-2	Benzene (including benzene from gasoline)
92-87-5	Benzidine
98-07-7	Benzotrichloride
100-44-7	Benzyl chloride
92-52-4	Biphenyl
117-81-7	Bis(2-ethylhexyl)phthalate (DEHP)
542-88-1	Bis(chloromethyl) ether
75-25-2	Bromoform
106-99-0	1,3-Butadiene
156-62-7	Calcium cyanamide
133-06-2	Captan
63-25-2	Carbaryl
75-15-0	Carbon disulfide
56-23-5	Carbon tetrachloride
463-58-1	Carbonyl sulfide
120-80-9	Catechol
133-90-4	Chloramben
57-74-9	Chlordane
7782-50-5	Chlorine
79-11-8	Chloroacetic acid
532-27-4	2-Chloroacetophenone
108-90-7	Chlorobenzene
510-15-6	Chlorobenzilate
67-66-3	Chloroform
107-30-2	Chloromethyl methyl ether
126-99-8	Chloroprene
1319-77-3	Cresol/Cresylic acid (mixed isomers)
95-48-7	o-Cresol
108-39-4	m-Cresol
106-44-5	p-Cresol
98-82-8	Cumene
N/A	2,4-D, salts and esters
72-55-9	DDE (1,1-dichloro-2,2-bis(p-chlorophenyl) ethylene)

Table 6-1 Section 112 Hazardous Air Pollutants (Continued)

Chemical abstracts service number	Pollutant
334-88-3	Diazomethane
132-64-9	Dibenzofuran
96-12-8	1,2-Dibromo-3-chloropropane
84-74-2	Dibutyl phthalate
106-46-7	1,4-Dichlorobenzene
91-94-1	3,3'-Dichlorobenzidine
111-44-4	Dichloroethyl ether (Bis[2-chloroethyl]ether)
542-75-6	1,3-Dichloropropene
62-73-7	Dichlorvos
111-42-2	Diethanolamine
64-67-5	Diethyl sulfate
119-90-4	3,3'-Dimethoxybenzidine
60-11-7	4-Dimethylaminoazobenzene
121-69-7	N,N-Dimethylaniline
119-93-7	3,3'-Dimethylbenzidine
79-44-7	Dimethylcarbamoyl chloride
68-12-2	N,N-Dimethylformamide
57-14-7	1,1-Dimethylhydrazine
131-11-3	Dimethyl phthalate
77-78-1	Dimethyl sulfate
N/A	4,6-Dinitro-o-cresol (including salts)
51-28-5	2,4-Dinitrophenol
121-14-2	2,4-Dinitrotoluene
123-91-1	1,4-Dioxane (1,4-Diethyleneoxide)
122-66-7	1,2-Diphenylhydrazine
106-89-8	Epichlorohydrin (1-Chloro-2,3-epoxypropane)
106-88-7	1,2-Epoxybutane
140-88-5	Ethyl acrylate
100-41-4	Ethylbenzene
51-79-6	Ethyl carbamate (Urethane)
75-00-3	Ethyl chloride (Chloroethane)
106-93-4	Ethylene dibromide (Dibromoethane)
107-06-2	Ethylene dichloride (1,2-Dichloroethane)
107-21-1	Ethylene glycol
151-56-4	Ethyleneimine (Aziridine)
75-21-8	Ethylene oxide
96-45-7	Ethylene thiourea
75-34-3	Ethylidene dichloride (1,1-Dichloroethane)
50-00-0	Formaldehyde
76-44-8	Heptachlor
118-74-1	Hexachlorobenzene

Table 6-1 Section 112 Hazardous Air Pollutants (Continued)

Chemical abstracts service number	Pollutant
87-68-3	Hexachlorobutadiene
N/A	1,2,3,4,5,6-Hexachlorocyclyhexane (all stereo isomers, including lindane)
77-47-4	Hexachlorocyclopentadiene
67-72-1	Hexachloroethane
822-06-0	Hexamethylene diisocyanate
680-31-9	Hexamethylphosphoramide
110-54-3	Hexane
302-01-2	Hydrazine
7647-01-0	Hydrochloric acid (Hydrogen chloride [gas only])
7664-39-3	Hydrogen fluoride (Hydrofluoric acid)
123-31-9	Hydroquinone
78-59-1	Isophorone
108-31-6	Maleic anhydride
67-56-1	Methanol
72-43-5	Methoxychlor
74-83-9	Methyl bromide (Bromomethane)
74-87-3	Methyl chloride (Chloromethane)
71-55-6	Methyl chloroform (1,1,1-Trichloroethane)
78-93-3	Methyl ethyl ketone (2-Butanone)
60-34-4	Methylhydrazine
74-88-4	Methyl iodide (Iodomethane)
108-10-1	Methyl isobutyl ketone (Hexone)
624-83-9	Methyl isocyanate
80-62-6	Methyl methacrylate
1634-04-4	Methyl tert-butyl ether
101-14-4	4,4'-Methylenebis(2-chloroaniline)
75-09-2	Methylene chloride (Dichloromethane)
101-68-8	4,4'-Methylenediphenyl diisocyanate (MDI)
101-77-9	4,4'-Methylenedianiline
91-20-3	Naphthalene
98-95-3	Nitrobenzene
92-93-3	4-Nitrobiphenyl
100-02-7	4-Nitrophenol
79-46-9	2-Nitropropane
684-93-5	N-Nitroso-N-methylurea
62-75-9	N-Nitrosodimethylamine
59-89-2	N-Nitrosomorpholine
56-38-2	Parathion
82-68-8	Pentachloronitrobenzene (Quintobenzene)

Table 6-1 Section 112 Hazardous Air Pollutants (Continued)

Chemical abstracts service number	Pollutant
87-86-5	Pentachlorophenol
108-95-2	Phenol
106-50-3	p-Phenylenediamine
75-44-5	Phosgene
7803-51-2	Phosphine
N/A	Phosphorus Compounds
85-44-9	Phthalic anhydride
1336-36-3	Polychlorinated biphenyls (Aroclors)
1120-71-4	1,3-Propane sultone
57-57-8	beta-Propiolactone
123-38-6	Propionaldehyde
114-26-1	Propoxur (Baygon)
78-87-5	Propylene dichloride (1,2-Dichloropropane)
75-56-9	Propylene oxide
75-55-8	1,2-Propylenimine (2-Methylaziridine)
91-22-5	Quinoline
106-51-4	Quinone (p-Benzoquinone)
100-42-5	Styrene
96-09-3	Styrene oxide
1746-01-6	2,3,7,8-Tetrachlorodibenzo-p-dioxin
79-34-5	1,1,2,2-Tetrachloroethane
127-18-4	Tetrachloroethylene (Perchloroethylene)
7550-45-0	Titanium tetrachloride
108-88-3	Toluene
95-80-7	Toluene-2,4-diamine
584-84-9	2,4-Toluene diisocyanate
95-53-4	o-Toluidine
8001-35-2	Toxaphene (chlorinated camphene)
120-82-1	1,2,4-Trichlorobenzene
79-00-5	1,1,2-Trichloroethane
79-01-6	Trichloroethylene
95-95-4	2,4,5-Trichlorophenol
88-06-2	2,4,6-Trichlorophenol
121-44-8	Triethylamine
1582-09-8	Trifluralin
540-84-1	2,2,4-Trimethylpentane
108-05-4	Vinyl acetate
593-60-2	Vinyl bromide
75-01-4	Vinyl chloride
75-35-4	Vinylidene chloride (1,1-Dichloroethylene)
1330-20-7	Xylenes (mixed isomers)
95-47-6	o-Xylene

Table 6-1 Section 112 Hazardous Air Pollutants (Continued)

Chemical abstracts service number	Pollutant
108-38-3	m-Xylene
106-42-3	p-Xylene
	Antimony Compounds
	Arsenic Compounds (inorganic including arsine)
	Beryllium Compounds
	Cadmium Compounds
	Chromium Compounds
	Cobalt Compounds
	Coke Oven Emissions
	Cyanide Compounds1
	Glycol ethers2
	Lead Compounds
	Manganese Compounds
	Mercury Compounds
	Fine mineral fibers3
	Nickel Compounds
	Polycyclic Organic Matter4
	Radionuclides (including radon)5
	Selenium Compounds

Note: For all listings above which contain the word "compounds" and for glycol ethers, the following applies: Unless otherwise specified, these listings are defined as including any unique chemical substance that contains the named chemical (i.e., antimony, arsenic, etc.) as part of that chemical's infrastructure.

1X'CN where X = H' or any other group where a formal dissociation may occur. For example, KCN or $Ca(CN)_2$.

2 (Under review)

Glycol Ether definition draft options:

Possible Correction to CAA 112(b)(1) footnote that would be consistant with OPPTS modified definition.

New OPPTS definition as published is:

R - $(OCH_2CH_2)n$ - OR' where:

n = 1, 2, or 3

R = alkyl C7 or less

or R = phenyl or alkyl

substituted phenyl

R' = H or alkyl C7 or less

or OR'= carboxylic acid ester, sulfate, phosphate, nitrate or sulfonate

CAA Glycol ether definition exactly as in the statute (errors included):

"Includes mono- and di ethers of ethylene glycol, diethylene glycol, and triethylene glycol R - $(OCH_2CH_2)_n$-OR' where

n = 1, 2, or 3

R = alkyl or aryl groups

Table 6-1 Section 112 Hazardous Air Pollutants (Continued)

$R' =$ R,H or groups which, when removed, yield glycol ethers with the structure R-(OCH2CH)n-OH. Polymers are excluded from the glycol category.

CAA Glycol ether definition with technical correction made.

(A 2 was left out of the last formula)
"Includes mono- and di- ethers of ethylene glycol, diethylene glycol, and triethylene glycol R-(OCH2CH2)$_n$ -OR' where

n = 1, 2, or 3

R = alkyl or aryl groups

R' = R, H, or groups which, when removed, yield glycol ethers with the structure: R-(OCH2CH2)$_n$-OH. Polymers are excluded from the glycol category.

3 (Under Review)

4 (Under Review)

5A type of atom which spontaneously undergoes radioactive decay.

of HAPs after application of air pollution controls. Unlike the definition of a source for New Source Review, HAP sources are not divided if they include operations with different SIC codes. All of the sources within a contiguous area are counted toward the 10- or 25-ton per year threshold. Air toxic requirements in individual states or local jurisdictions may be lower than these thresholds.

Title III also specified control level requirements for HAP sources termed Maximum Achievable Control Technologies (MACT). MACT was defined in Chapter 3, but for completeness, the definition is repeated here. As specified in Title III of the Clean Air Act Amendments of 1990, MACT is:

> . . . the maximum degree of reduction in emissions of the HAPs subject to this section (including prohibitions on such emissions, where achievable) that the Administrator, taking into consideration the cost of achieving such emission reduction, and any nonair quality health and environmental impacts and energy requirements, determines is achievable for new or existing sources in the category or subcategory to which such emission standard applies . . .[1]

In addition to the general definition of MACT, the CAAA also specified a "MACT floor," the minimum level of control that could be determined to be MACT. This MACT floor was specified as:

> The maximum degree of reduction that is deemed achievable for new sources...shall not be less stringent than the emission control that is achieved in practice by the best controlled similar source. Emission standards for existing sources may be less stringent than for new sources but shall not be less stringent than the average emissions limitation achieved by the best performing 12% of existing sources (excluding those which have recently complied with LAER).[2]

Instead of identifying control technologies for each HAP, Title III specified that HAP controls would be defined by source category. Unlike BACT, MACT is determined for a category or subcategory of sources as a whole and not on a case-by-case basis. Based on the above definition, MACT control levels are likely to be 90% or greater for organics, and 95% or greater for particulates.

The determination of MACT for each source category will be important for applicants planning new or modified sources. If a MACT standard has already been promulgated, it will include requirements that must be met by new or modified sources. The schedule for promulgation of MACTs for different source categories is shown in Figure 6-1. For MACTs that have already been promulgated, *Federal Register* citations to the standards are shown in Figure 6-2.[3]

If no MACT standard has yet been promulgated for the source being permitted, the permitting agency will need to make a MACT determination. The applicant may submit information to support a proposed MACT demonstrating that it is equivalent to the control on the best controlled source currently operating.[4]

Unlike BACT, which must be in place prior to startup of a new or modified source, MACT must be in place at startup only for new sources. If the source is a modification, the applicant can request an extension of up to 3 years before MACT must be in place.

III. NEW OR MODIFIED HAP SOURCE PERMITTING

In addition to specifying control technology requirements for new and existing sources of HAP, the 1990 Amendments specified two requirements for new or modified HAP source permitting. These were offsets and control technologies.[14] Although mandated by the Amendments to promulgate rules for offset and control technologies within 18 months after passage of the Amendments, no rules have been promulgated. EPA has indicated that its greater priority is in establishing MACT standards for all of the categories required by the Amendments so that those standards can be applied to any new HAP sources. Until a challenge is made to this policy, HAP sources will only be subject to the applicable MACT(s), with the stringency of controls determined by whether the source is defined, in the applicability for specific MACT(s), as new or existing.

IV. HEALTH RISK ASSESSMENTS FOR PERMITTING

In many jurisdictions, health risk assessments are used to determine whether impacts from hazardous or toxic air pollutant emissions are acceptable. The tool that is used is called a multipathway health risk assessment (HRA). It is a quantitative evaluation of the potential for adverse health effects due to exposure to toxic substances in the environment. Health risk assessments have been used to assess a variety of health related effects from pollutants, including water-borne and solid residual pollutants, but its most common application is to assess the impacts of releases to the air. The requirements for conducting a multipathway HRA have been prescribed both by EPA[7,8] and by several states.[9] The objectives of the HRA are to determine the risk of cancer due to pollutants from a proposed

FUEL COMBUSTION

Category	Promulgation
Engine Test Facilities	11/15/2000
Industrial Boilers	11/15/2000
Institutional/Commercial Boilers	11/15/2000
Process Heaters	11/15/2000
Stationary Internal Combustion Engines	11/15/2000
Stationary Turbines	11/15/2000

NON-FERROUS METAL PROCESSING

Category	Promulgation
Primary Aluminum Production	11/15/1997
Secondary Aluminum Production	11/15/1997
Primary Copper Smelting	11/15/1997
Primary Lead Smelting	11/15/1997
Secondary Lead Smelting	11/15/1994
Lead Acid Battery Manufacturing	11/15/2000
Primary Magnesium Refining	11/15/2000

FERROUS METALS PROCESSING

Category	Promulgation
Coke By-Product Plants	11/15/2000
Coke Ovens: Charging, Top Side, and Door Leaks	12/31/1992
Coke Ovens: Pushing, Quenching, and Battery Stacks	11/15/2000
Ferroalloys Production	11/15/1997
Integrated Iron and Steel Manufacturing	11/15/2000
Non-Stainless Steel Manufacturing - Electric Arc Furnace (EAF) Operation	11/15/1997
Stainless Steel Manufacturing - Electric Arc Furnace (EAF) Operation	11/15/1997
Iron Foundries	11/15/2000
Steel Foundries	11/15/2000
Steel Pickling - HCl Process	11/15/1997

MINERAL PRODUCTS PROCESSING

Category	Promulgation
Alumina Processing	11/15/2000
Asphalt Concrete Manufacturing	11/15/2000
Asphalt Processing	11/15/2000
Asphalt Roofing Manufacturing	11/15/2000
Asphalt/Coal Tar Application - Metal Pipes	11/15/2000
Chromium Refractories Production	11/15/2000
Clay Products Manufacturing	11/15/2000
Lime Manufacturing	11/15/2000
Mineral Wool Production	11/15/1997
Portland Cement Manufacturing	11/15/1997
Taconite Iron Ore Processing	11/15/2000
Wool Fiberglass Manufacturing	11/15/1997

PETROLEUM AND NATURAL GAS PRODUCTION AND REFINING

Category	Promulgation
Oil and Natural Gas Production	11/15/1997
Petroleum Refineries - Catalytic Cracking (Fluid and other) Units, Catalytic Reforming Units, and Sulfur Plant Units	11/15/1997
Petroleum Refineries - Other Sources Not Distinctly Listed	11/15/1994

PRODUCTION OF ORGANIC CHEMICALS

Category	Promulgation
Synthetic Organic Chemical Manufacturing	11/15/1992

SURFACE COATING PROCESSES

Category	Promulgation
Aerospace Industries	11/15/1994
Auto and Light Duty Truck (Surface Coating)	11/15/2000
Flat Wood Paneling (Surface Coating)	11/15/2000
Large Appliance (Surface Coating)	11/15/2000
Magnetic Tapes (Surface Coating)	11/15/1994
Manufacture of Paints, Coatings, and Adhesives	11/15/2000
Metal Can (Surface Coating)	11/15/2000
Metal Coil (Surface Coating)	11/15/2000
Metal Furniture (Surface Coating)	11/15/2000
Miscellaneous Metal Parts and Products (Surface Coating)	11/15/2000
Paper and Other Webs (Surface Coating)	11/15/2000
Plastic Parts and Products (Surface Coating)	11/15/2000
Printing, Coating, and Dyeing of Fabrics	11/15/2000
Printing/Publishing (Surface Coating)	11/15/1994
Shipbuilding and Ship Repair (Surface Coating)	11/15/1994
Wood Furniture (Surface Coating)	11/15/1994

MISCELLANEOUS PROCESSES

Category	Promulgation
Aerosol Can-Filling Facilities	11/15/2000
Benzyltrimethylammonium Chloride Production	11/15/2000
Butadiene Dimers Production	11/15/1997
Carbonyl Sulfide Production	11/15/2000
Chelating Agents Production	11/15/200
Chlorinated Paraffins Production	11/15/2000
Chromic Acid Anodizing	11/15/1994
Commercial Dry Cleaning (Perchloroethylene) - Transfer Machines	11/15/1992
Commercial Sterilization Facilities	11/15/1994
Decorative Chromium Electroplating	11/15/1994
Dodecanedioic Acid Production	11/15/2000
Dry Cleaning (Petroleum Solvent)	11/15/2000
Ethylidene Norbornene Production	11/15/2000
Explosives Production	11/15/2000
Halogenated Solvent Cleaners	11/15/1994
Hard Chromium Electroplating	11/15/1994
Hydrazine Production	11/15/2000
Industrial Cleaning (Perchloroethylene) - Dry-to-dry machines	11/15/1992
Industrial Dry Cleaning (Perchloroethylene) - Transfer Machines	11/15/1992
Industrial Process Cooling Tower	11/15/1994
OBPA/1,3-Diisocyanate Production	11/15/2000
Paint Stripper Users	11/15/2000
Photographic Chemicals Production	11/15/2000
Phthalate Plasticizers Production	11/15/2000
Plywood/Particle Board Manufacturing	11/15/2000
Polyether Polyols Production	11/15/1997
Pulp and Paper Production	11/15/1997
Rocket Engine Test Firing	11/15/2000
Rubber Chemicals Manufacturing	11/15/2000
Semiconductor Manufacturing	11/15/2000
Symmetrical Tetrachloropyridine Production	11/15/2000
Tire Production	11/15/2000
Wood Treatment	11/15/1997

Figure 6-1 Schedule of due dates for promulgation of MACTs. (Source: 58 Federal Register 63941 (12/03/93).)

LIQUIDS DISTRIBUTION

Category	Promulgation
Gasoline Distribution (Stage 1)	11/15/1994
Organic Liquids Distribution (Non-Gasoline)	01/15/2000

CATEGORIES OF AREA SOURCES

Category	Promulgation
Asbestos Processing	11/15/1994
Chromic Acid Anodizing	11/15/1994
Commercial Dry Cleaning (Perchloroethylene) - Dry-to-Dry Machines	11/15/1992
Commercial Dry Cleaning (Perchloroethylene)- Transfer Machines	11/15/1992
Commercial Sterilization Facilities	11/15/1994
Decorative Chromium Electroplating	11/15/1994
Halogenated Solvent Cleaners	11/15/1994
Hard Chromium Electroplating	11/15/1994

WASTE TREATMENT AND DISPOSAL

Category	Promulgation
Hazardous Waste Incineration	11/15/2000
Municipal Landfills	11/15/2000
Publicly Owned Treatment Works (POTW) Emissions	11/15/1995
Sewage Sludge Incineration	11/15/2000
Site Remediation	11/15/2000
Solid Waste Treatment, Storage and Disposal Facilities (TSDF)	11/15/1994

AGRICULTURAL CHEMICALS PRODUCTION

Category	Promulgation
4-Chloro-2-Methylphenoxyacetic Acid Production	11/15/1997
2,4-D Salts and Esters Production	11/15/1997
4,6-Dinitro-o-Cresol Production	11/15/1997
Captafol Production	11/15/1997
Captan Production	11/15/1997
Chloroneb Production	11/15/1997
Chlorothalonil Production	11/15/1997
Dacthal (tm) Production	11/15/1997
Sodium Pentachlorophenate Production	11/15/1997
Tordon (tm) Acid Production	11/15/1997

FIBERS PRODUCTION PROCESSES

Category	Promulgation
Acrylic Fibers/Modacrylic Fibers Production	11/15/1997
Rayon Production	11/15/2000
Spandex Production	11/15/2000

FOOD AND AGRICULTURE PROCESSES

Category	Promulgation
Baker's Yeast Manufacturing	11/15/2000
Cellulose Food Casing Manufacturing	11/15/2000
Vegetable Oil Production	11/15/2000

PHARMACEUTICAL PRODUCTION PROCESSES

Category	Promulgation
Pharmaceuticals Production	11/15/1997

POLYMERS AND RESINS PRODUCTION

Category	Promulgation
Acetal Resins Production	11/15/1997
Acrylonitrile-Butadiene-Styrene Production	11/15/1994
Alkyd Resins Production	11/15/2000
Amino Resins Production	11/15/1997
Boat Manufacturing	11/15/2000
Butadiene-Furfural Cotrimer (R-11)	11/15/2000
Butyl Rubber Production	11/15/1994
Carboxymethylcellulose Production	11/15/2000
Cellophane Production	11/15/2000
Cellulose Ethers Production	11/15/2000
Epichlorohydrin Elastomers Production	11/15/1994
Epoxy Resins Production	11/15/1994
Ethylene-Propylene Rubber Production	11/15/1994
Flexible Polyurethane Foam Production	11/15/1997
Hypalon (tm) Production	11/15/1994
Maleic Anhydride Copolymers Production	11/15/2000
Methylcellulose Production	11/15/2000
Methyl Methacrylate-Acrylonitrile Butadiene-Styrene Production	11/15/1994
Methyl Methacrylate-Butadiene-Styrene Terpolymers Production	11/15/1994
Neoprene Production	11/15/1994
Nitrile Butadiene Rubber Production	11/15/1994
Non-Nylon Polyamides Production	11/15/1994
Nylon 6 Production	11/15/1997
Phenolic Resins Production	11/15/1997
Polybutadiene Rubber Production	11/15/1994
Polycarbonates Production	11/15/1997
Polyester Resins Production	11/15/2000
Polyethylene Terephthalate Production	11/15/1994
Polymerized Vinylidene Chloride Production	11/15/2000
Polymethyl Methacrylate Resins Production	11/15/2000
Polystyrene Production	11/15/1994
Polysulfide Rubber Production	11/15/1994
Polyvinyl Acetate Emulsions Production	11/15/2000
Polyvinyl Alcohol Production	11/15/2000
Polyvinyl Butyral Production	11/15/2000
Polyvinyl Chloride and Copolymers Production	11/15/2000
Reinforced Plastic Composites Production	11/15/1997
Styrene-Acrylonitrile Production	11/15/1994
Styrene-Butadiene Rubber and Latex Production	11/15/1994

PRODUCTION OF INORGANIC CHEMICALS

Category	Promulgation
Ammonium Sulfate Production - Caprolactam By-Product Plants	11/15/2000
Antimony Oxides Manufacturing	11/15/2000
Chlorine Production	11/15/1997
Chromium Chemicals Manufacturing	11/15/1997
Cyanuric Chloride Production	11/15/1997
Fume Silica Production	11/15/2000
Hydrochloric Acid Production	11/15/2000
Hydrogen Cyanide Production	11/15/1997
Hydrogen Fluoride Production	11/15/2000
Phosphate Fertilizers Production	11/15/2000
Phosphoric Acid Manufacturing	11/15/2000
Quaternary Ammonium Compounds Production	11/15/2000
Sodium Cyanide Production	11/15/1997
Uranium Hexafluoride Production	11/15/2000

Figure 6-1 *(Continued).*

facility, evaluate the potential for chronic noncancer effects, and evaluate the potential for acute noncancer effects.

Risk assessment is a multidisciplinary analysis. The data from different disciplines required and the analyses using different methodologies cannot be developed by one discipline. Once the protocol for a HRA has been developed, and the first one or several completed, persons outside of the primary disciplinary areas required can execute the calculations. However, methodological advancements are almost always accomplished by interdisciplinary teams.

NESHAP (MACT STANDARDS)
(Proposed or Promulgated Rules only)

Source Category	40 CFR Part 63 Subpart	Proposal	Promulgation	Sources Covered
Standards Scheduled for Promulgation by 1992[a]				
Coke Ovens: Charging, topside and door leaks	L	12/04/92 (57 FR 57534)	10/27/93 (58 FR 57898) 01/13/94 (59 FR 1922)-C[b]	Major
Dry Cleaning (all listed categories)	M	12/09/91 (56 FR 64382) 03/31/93 (58 FR 16808)-P[b]	09/22/93 (58 FR 49354) 12/20/93 (58 FR 66287)-A[b] 12/13/95 (60 FR 64002)-A[b]	Major and Area
Synthetic Organic Chemical Manufacturing (a/k/a HON)	F · I	12/31/92 (57 FR 62608)[c]	04/22/94 (59 FR 19402)[c]	Major
Standards Scheduled for Promulgation by 1994[a]				
Aerospace Industry	GG	06/06/94 (59 FR 29216) 08/01/94 (59 FR 38949)-P[b] 11/22/94 (59 FR 60101)-P[b]	09/01/95 (60 FR 45948)	Major
Asbestos		Proposed delisting: 01/24/95 (60 FR 4624) 07/28/95 (60 FR 38725)-C[b]	Final delisting: 11/30/95 (60 FR 61550)	Delisted
Chromium Electroplating -Chromic Acid Anodizing -Decorative Chromium Electroplating -Hard Chromium Electroplating	N	12/16/93 (58 FR 65768) 03/04/94 (59 FR 10352)-P[b]	01/25/95 (60 FR 4948) 05/24/95 (60 FR 27598)-C[b] 06/27/95 (60 FR 33122)-C[b] 12/13/95 (60 FR 64002)-A[b]	Major and Area
Commercial (Ethylene Oxide) Sterilizers	O	03/07/94 (59 FR 10591)	12/06/94 (59 FR 62585) 12/13/95 (60 FR 64002)-A[b]	Major and Area
Gasoline Distribution/Marketing - Stage I	R	02/08/94 (59 FR 5868) 03/04/94 (59 FR 10461)-C[b] 08/19/94 (59 FR 42788)-P[b]	12/14/94 (59 FR 64303) 02/08/95 (60 FR 7627)-C[b] 06/26/95 (60 FR 32912)-C[b] 11/07/95 (60 FR 56133)-A[b] 12/08/95 (60 FR 62991)-S[b] 02/29/96 (61 FR 7718)-A[b]	Major
Halogenated Solvent Cleaning Machines	T	11/29/93 (58 FR 62566)	12/02/94 (59 FR 61801) 12/30/94 (59 FR 67750)-C[b] 06/05/95 (60 FR 29484)-C[b]	Major and Area
Industrial Process Cooling Towers	Q	08/12/93 (58 FR 43028)	09/08/94 (59 FR 46339)	Major
Magnetic Tape	EE	03/11/94 (59 FR 11662)	12/15/94 (59 FR 64580)	Major
Offsite Waste Treatment Operations	DD	10/13/94 (59 FR 51913) 12/21/94 (59 FR 65744)-P[b]	Court ordered: 05/10/96	Major
Petroleum Refinery - Other Sources Not Distinctly Listed	CC	07/15/94 (59 FR 36130)	08/18/95 (60 FR 43244) 09/27/95 (60 FR 49976)-C[b]	Major

Figure 6-2 *Federal Register* citations for promulgation of MACTs. (Source: Susan J. Miller, Radian International LLC.)

Risk assessment is composed of four main components. These are hazard identification, exposure assessment, dose-response assessment, and risk characterization. Each of these will be discussed below.

A. Hazard Identification

The hazardous or toxic air pollutants that are expected to be emitted by a proposed project are the hazard. Both the speciation of air toxics and the rate of emissions must be determined to proceed with the remaining steps of the HRA. The approaches used to estimate toxics emissions are described earlier in this chapter and in Chapter 2.

Often, to simplify calculations, a surrogate pollutant is selected as the identified hazard. An example is when benzo-a-pyrene (BAP) is selected as the surrogate for polyaromatic hydrocarbons (PAH) emitted from low-temperature combustion processes, such as wood-burning stoves or fireplaces. BAP is chosen because it has a higher potential for causing cancer than any of the other PAHs,

Petroleum Refinery - Other Sources Not Distinctly Listed	CC	07/15/94 (59 FR 36130)	08/18/95 (60 FR 43244) 09/27/95 (60 FR 49976)-C[b]	Major
Polymers and Resins I ·Butyl Rubber ·Epichlorohydrin Elastomers ·Ethylene Propylene Rubber ·Hypalon™ ·Neoprene ·Nitrile Butadiene Rubber ·Polybutadiene Rubber ·Polysulfide Rubber ·Styrene-Butadiene Rubber and Latex	U	06/12/95 (60 FR 30801)	Promulgation expected by 07/15/96	Major
Polymers and Resins II ·Epoxy Resins Production ·Non-nylon Polyamides Production	W	05/16/94 (59 FR 25387)	03/08/95 (60 FR 12670)	Major
Polymers and Resins IV ·Acrylonitrile-Butadiene-Styrene ·Methyl Methacrylate Acrylonitrile-Butadiene-Styrene ·Methyl Methacrylate Butadiene-Styrene Terpolymers ·Polystyrene ·Styrene Acrylonitrile ·Polyethylene Terephthalate	V	03/29/95 (60 FR 16090)	Court ordered: 05/15/96	Major
Printing and Publishing (Surface Coating)	KK	03/14/95 (60 FR 13664) 04/03/95 (60 FR 16920)-C[b]	Court ordered: 05/15/96	Major
Secondary Lead Smelters	X	06/09/94 (59 FR 29750) 04/19/95 (60 FR 19556)-P[b]	06/23/95 (60 FR 32587) 12/13/95 (60 FR 64002)-A[b]	Major and area
Shipbuilding (Surface Coating)	II	12/06/94 (59 FR 62681)	12/15/95 (60 FR 64330)	Major
Wood Furniture (Surface Coating)	JJ	12/06/94 (59 FR 62652) 02/22/95 (60 FR 9812)-P[b]	12/07/95 (60 FR 62930)	Major
Standards Scheduled for Promulgation by 1997[a] **Pulp and Paper (non-combustion)**	S	12/17/93 (58 FR 66078) 02/22/95 (60 FR 9813)-D[b] 07/05/95 (60 FR 34938)-D[b] 03/08/96 (61 FR 9383)-D,P[b]	Promulgation expected: 08/96	Major
Standards Scheduled for Promulgation by 2000[a] **Marine Vessels**	Y	05/13/94 (59 FR 25004) 08/31/94 (59 FR 44955)-P[b] 03/08/96 (60 FR 12723)-P[b]	09/19/95 (60 FR 48388)	Major

led" date refers to schedule established under Section 112(e) (12/03/93, 58 FR 63941)
ndment   C = Correction   P = Additional public comment   D = Notice of Data Availability   S = Stay of Compliance
iental HON / SOCMI notices: 02/24/93 (58 FR 11201)-P;   02/26/93 (58 FR 11667);   10/15/93 (58 FR 53478)-P;
i (59 FR 11018) (EPA position on MACT "floor")   06/06/94 (59 FR 29196)   09/20/94 (59 FR 48175)-C
i (59 FR 53359)-S   10/24/94 (59 FR 53392, 53395)-A   10/28/94 (59 FR 54131)-S   10/28/94 (59 FR 54154)-A
i (60 FR 5320)-S   04/10/95 (60 FR 18020, 18026, 18071, 18078)-A,C (4 notices)   12/12/95 (60 FR 63624)-A
i (61 FR 7716, 7761)-AC (2 notices)

Figure 6-2 *(Continued).*

and using it as a surrogate simplifies the calculations, requires less data, and assures that the HRA will be conservative — that is, it will overestimate the health risk. Note that this simplification comes at a cost. A higher risk of cancer will be estimated using this surrogate approach than to proceed in the calculations with detailed speciation of the different PAH compounds. On the other hand, acquiring additional speciated data may be expensive and time consuming.

B. Exposure Assessment

The exposure assessment includes transport modeling, environmental fate, and exposure assessments. Dispersion models are used to estimate concentrations in ambient air and deposition rates for pollutant particulates to fall to the ground and into any water bodies within the area of impacts. Dispersion modeling was discussed in Chapter 5.

Fate models are used to predict the behavior of pollutants in soil and water. These models describe how a pollutant may degrade in water, or may concentrate in some species of fish.

Other models are used to estimate the concentrations in soil, and the uptake of pollutants through roots into plants. Additional pathways track pollutants as they accumulate in fruits and vegetables, or in grains or grasses, which then provide a pathway to beef, cow's milk, and chickens. Simple screening models are available to predict the fate of pollutants. However, as noted above, simplicity comes at the cost of yielding a more conservative estimate of health risk impacts.

C. Exposure Estimate

The exposure estimate synthesizes exposure through direct inhalation of toxic pollutants, through ingestion, and through dermal contact. Inhalation may be either of gaseous pollutants or small particulates that are absorbed in the small passages in the lungs. As discussed in Chapter 1, only small particulates, usually less than 1 or 2 μm in diameter, can penetrate far enough into the lungs to cause health effects.

Ingestion can be of anything that enters the mouth. This includes plants, dairy products, beef, poultry, pork, fish, water, mother's milk, or even dirt. One of the pathways that is considered is for young children exhibiting *pica,* an appetite or craving for unnatural food such as dirt.

Dermal contact can provide a pathway for absorption of polluted water, soil, or dust. This pathway is usually a modest contributor to risk because of the substantially impervious character of skin.

Not all pathways are considered in a risk assessment. The pathways that are important will depend upon the land use in the vicinity of the emission source and the life-style of persons exposed. For example, the HRA for a new emission source in a farming community would need to consider all of the soil and plant uptake pathways, whereas an urban source would provide negligible pathway through plant uptake. In another example, an area in which there are no dairy herds would not need to include dairy pathways.

Most risk assessments require the evaluation of risk to the maximum exposed individual (MEI). The MEI is assumed to receive continuous exposure for a lifetime (assumed to be 70 years) at the point of maximum impact from the combination of all pathways. Thus the MEI would live where the highest air pollutant concentration occurs, would drink water from a pond on the property, have a home garden from which he obtained all his food, and have cattle and chickens from which he obtained all dairy and meat products.

The use of the MEI is a conservative approach to estimating risk to actual populations. Most people live in one place an average of 6.7 years, not 70 years. Few people raise all the food they eat in their own gardens, and most people leave their homes for several hours each day to go to school or work. Thus the MEI ignores microenvironment and mobility, which, if included, would lower health risk estimates.

The end result of the exposure assessment is an estimate of total daily intake of pollutants from inhalation, ingestion, and dermal contact. The intake is usually expressed in units of milligrams of each toxic substance per kilogram of body weight per day of exposure.

D. Dose-Response Assessment

The dose-response assessment provides toxicological factors that predict the likelihood of cancer or other health effects occurring as a result of exposure. Cancer potency or slope and unit risk factors (URF) predict the likelihood of cancer. Reference dose (RfD) and chronic and acute acceptable exposure levels (AELs) predict the likelihood of noncarcinogenic health effects from chronic or acute doses. EPA is the primary developer of dose-response factors based upon toxicological or epidemiological data.

Cancer potency factors, also known as slope factors, are developed assuming a zero threshold. That is, any dose level is assumed to produce some risk of contracting cancer. The most common model used to develop cancer potency estimates is the multistage linear extrapolation. It predicts a tumor incidence percentage based upon dose, as shown in Figure 6-3. The extrapolation model predicts a cancer response at very low doses. The doses due to environmental exposure may be from 1 million to 1 billion times smaller than the bioassay dose level at which tumors are observed in laboratory animals or through epidemiological studies of occupational exposures. This model extrapolates the bioassay dose to smaller doses by assuming that the threshold for cancer occurrence is zero, and any dose larger than zero results in some likelihood of cancer occurring.

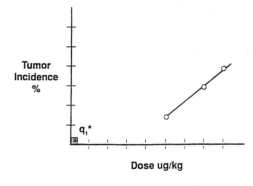

Figure 6-3
Extrapolation models predict a cancer response at very low doses. (Source: Michael T. Alberts, Radian International LLC.)

EPA reference dose (RfD) is a method for addressing chronic health effects from pollutants. The dose of a chronic pollutant is compared to the RfD. If the dose is equal to or greater than the RfD, the health effects due to this pollutant are perceived to be significant.

To determine values for the RfD, health effects data for the pollutant from animal studies are usually used. The process starts with the lowest available no observable effects level (NOEL) and reduces this level by several uncertainty/safety factors to obtain an RfD. The first uncertainty is due to the extrapolation of animal results to humans. The uncertainty factor from animal to human application is a factor of 10. The second uncertainty is going from an average human to a sensitive human. Health effect evaluation criteria are intended to protect the most sensitive population, not just the average individual. This is a second factor of 10. The third uncertainty is extrapolating the data from subchronic to chronic effects, another factor of 10. The fourth uncertainty is if the data are for lowest observable effects instead of no observable effects. If the RfD is based upon lowest observable effects, another factor of 10 is added to assure that there will be no observable effects. Finally, a professional assessment is done to evaluate the quality and extent of the data available. This assessment results in a modifying factor between 0 and 10. When all of these factors are combined, the NOEL can be reduced by between 100 and 100,000 times. The RfD is the result of this uncertainty analysis. Usually, the RD is 1/1000 of the NOEL.

E. Risk Characterization

Risk characterization provides a quantitative estimate of health effects from pollutants. The risk characterization usually includes lifetime cancer risk, cancer incidence or burden, acute noncancer effects, and chronic noncancer effects.

1. Cancer Risk

The cancer risk is the product of the exposure and cancer potency,

$$\text{Risk} = \left(\text{Exposure}\left[\frac{\text{mg}}{\text{kg}-\text{day}}\right]\right)\left(\text{SF}\left[\frac{\text{kg}-\text{day}}{\text{mg}}\right]\right) \qquad (6\text{-}1)$$

where exposure is determined from the exposure estimate and SF is the slope factor determined from laboratory animal or epidemiological studies.

Cancer risk may also be calculated for an exposure that is only due to inhalation by multiplying the pollutant concentration times a unit risk factor,

$$\text{Risk} = \left(\text{Concentration}\left[\frac{\mu\text{g}}{\text{m}^3}\right]\right)\left(\text{URF}\left[\frac{\text{m}^3}{\mu\text{g}}\right]\right) \qquad (6\text{-}2)$$

The unit risk factor (URF) is the product of the cancer potency and body weight (assumed to be 70 kg) divided by the human inhalation rate 20 m^3/day). Although most people do not weigh 70 kg, most people also do not breathe 20 cubic meters of air per day, so these two conservative numbers balance each other. Values for URF for different toxic air pollutants are shown in Table 6-2.

Risk is calculated for each pollutant emitted from a source or facility. The total risk is the sum of the risk from each pollutant and each pathway,

$$\text{Total risk} = \text{risk}_a + \text{risk}_b + \text{risk}_c + \ldots + \text{risk}_n \qquad (6\text{-}3)$$

where a through n are the pollutants that contribute to risk.

This is the approach recommended by EPA,[10] and it assumes that there are neither synergistic nor antagonistic effects that occur due to the presence of several pollutants. Although this assumption is perceived by some to be not health

Table 6-2   Unit Risk Factors for Some Carcinogenic Air Pollutants

Substance	Unit risk $(\mu g/m^3)^{-1}$	Reference
Acetaldehyde	2.2E-6	IRIS
Acrylamide	1.3E-3	IRIS/QEHHA-RCHAS
Acrylonitrile	2.9E-4	QEHHA-RCHAS
Arsenic	3.3E-3	QEHHA-ATES/ARB
Arsenic compounds (inorganic)	3.3E-3	QEHHA-ATES/ARB
Asbestos	[1.9E-4/100 fibers/m³][a]	QEHHA-ATES/ARB
Benzene	2.9E-5	QEHHA-RCHAS
Benzidene (and its salts)	1.4E-1	QEHHA-RCHAS
Beryllium	2.4E-3	IRIS
Bis(chloromethyl)ether	1.3E-2	QEHHA-RCHAS
1,3-Butadiene	2.8E-4	IRIS
Cadmium	4.2E-3	QEHHA-RCHAS
Cadmium compounds	4.2E-3	QEHHA-ATES/ARB
Carbon tetrachloride	4.2E-5	QEHHA-RCHAS, ATES
Chlorinated dibenzo-p-dioxins (as 2,3,7,8-equivalent)	3.8E+1	QEHHA-RCHAS, ATES/ARB
Chlorinated dibenzofurans (as 2,3,7,8-equivalents)	3.8E+1	QEHHA-RCHAS, ATES/ARB
Chloroform	5.3E-6	QEHHA-ATES/ARB
Chlorophenols		
Pentachlorophenol	4.6E-6	
2,4,6-Trichlorophenol	2.0E-5	QEHHA-RCHAS
Chloroprene	1.3E-7	QEHHA-RCHAS
Chromium (hexavalent)	1.4E-1	QEHHA-RCHAS
Coke oven emissions	6.2E-4	IRIS
1,2-Dibromo-3-chloropropane (DBCP)	2.0E-3	QEHHA-RCHAS
p-Dichlorobenzene (1,4-Dichlorobenzene)	1.1E-5	QEHHA-RCHAS
3,3'-Dichlorobenzidene	3.4E-4	QEHHA-RCHAS
Di(2-ethylhexyl)phthalate (DEHP)	2.4E-6	QEHHA-RCHAS
1,4-Dioxane	7.7E-6	QEHHA-RCHAS
Epichlorohydrin	2.3E-5	QEHHA-RCHAS
Ethylene dibromide (1,2-Dibromoethane)	7.1E-5	QEHHA-RCHAS, ATES/ARB
Ethylene dichloride (1,2-Dichloroethane)	2.0E-5	QEHHA-RCHAS, ATES/ARB
Ethylene oxide	8.8E-5	QEHHA-ATES/ARB
Formaldehyde	1.3E-5	IRIS
Gasoline vapors	1.2E-6[3.5E-3 ppm⁻¹]	QEHHA-RCHAS
Hexachlorobenzene	5.1E-4	QEHHA-RCHAS
Hexachlorocyclohexanes	1.1E-3	QEHHA-RCHAS
Hydrazine	4.9E-3	IRIS
Methylene chloride (Dichloromethane)	1.0E-6	QEHHA-ATES/ARB
Nickel and nickel compounds	2.6E-4	QEHHA-ATES/ARB
N-Nitrosodiethylamine	1.0E-2	QEHHA-RCHAS
N-Nitrosodimethylamine	4.6E-3	QEHHA-RCHAS
p-Nitrosodiphenylamine	2.6E-6	QEHHA-RCHAS
N-Nitrosodi-n-butylamine	3.1E-3	QEHHA-RCHAS
N-Nitrosomethylethylamine	6.3E-3	IRIS/QEHHA-RCHAS
N-Nitrosodi-n-propylamine	2.0E-3	QEHHA-RCHAS

Table 6-2 Unit Risk Factors for Some Carcinogenic Air Pollutants (Continued)

Substance	Unit risk $(\mu g/m^3)^{-1}$	Reference
N-Nitrosopyrrolidine	6.0E-4	IRIS/QEHHA-RCHAS
PCBs (Polychlorinated biphenyls)	1.4E-3	QEHHA-RCHAS
PAHs (Polycyclic aromatic hydrocarbons) including, but not limited to:		
Benz[a]anthracene	1.7E-3	Ref. 1
Benzo[b]fluoranthene	1.7E-3	Ref. 1
Benzo[k]fluoranthene	1.7E-3	Ref. 1
Benzo[a]pyrene	1.7E-3	Ref. 1
Dibenz[a,h]anthracene	1.7E-3	Ref. 1
Indeno[1,2,3-cd]pyrene	1.7E-3	Ref. 1
Perchloroethylene (Tetrachloroethylene)	5.8E-7	EPA/Ref. 2,3
Propylene oxide	3.7E-6	IRIS
Trichloroethylene	2.0E-6	QEHHA-ATES/ARB
Urethane	2.9E-4	QEHHA-RCHAS
Vinyl chloride	7.8E-5	QEHHA-ATES/ARB

[a] A conversion factor of 100 fibers/0.003 μg can be multiplied by a receptor concentration of asbestos expressed in terms of micrograms per cubic meter $(\mu g/m^3)$ to yield fibers/m³ (EPA, 1985. Airborne Asbestos Health Risk Assessment Update). Unless other information necessary to estimate the concentration (fibers/m³) of asbestos at receptors of interest is available, the use of the aforementioned conversion factor is an option.
References:
IRIS refers to the U.S. Environmental Protection Agency's Integrated Risk Information System Database.
QEHHA refers to the California EPA Office of Environmental Health Hazard Assessment.
QEHHA-ATES/ARB refers to reports by the California Office of Environmental Health Hazard Assessment, Air Toxicology Epidemiology Section and the California Air Resources Board prepared in the process of identifying the material as a Toxic Air Contaminant.
QEHHA-RCHAS refers to reports prepared by the California Office of Environmental Health Hazard Assessment, Reproductive and Cancer Hazard Assessment Section as part of the implementation of the Safe Drinking Water and Toxic Enforcement Act of 1988 (Proposition 65).
1. EPA, 1984. *Health Effects Document for Benzo[a]pyrene.* EPA/540/1-86/022, September, 1984. NTIS PB86-134335.
2. EPA, 1985. *Health Assessment Document for Tetrachloroethylene (Perchloroethylene).* Final Report. EPA-600/8-82/005F, July, 1985, [oral potency estimate].
3. Federal Register 52880-52884, 9/26/85 [inhalation potency estimate].
Source: California Air Pollution Control Officers Association (CAPCOA) "Air Toxics 'Hot Spots' Program Risk Assessment Guidelines," January, 1992.

conservative, there are very little data on synergistic or antagonistic effects of two pollutants, let alone several, and all of the data are at much higher doses than those that occur in environmental exposures.

2. Cancer Burden

Cancer burden is the number of additional cancers in the population exposed to a pollutant or pollutants. It is determined by taking the risk times the population,

$$\text{Burden} = (\text{Risk})(\text{Population}) \qquad (6\text{-}4)$$

Since the concentration of pollutants, and consequently the risk due to pollutants decreases with distance from a source, not all populations experience the same risk level. Consequently, to determine burden, the risk is calculated at the centroid of census tracts around the facility. The burden for each census tract is calculated as that risk times the population of the tract. This is then summed for all of the census tracts in the vicinity of the facility for which the centroid risk is greater than 1 in 10 million,

$$\text{Burden} = \sum_{j=1}^{n} \text{Risk}_j(\text{Population}_j) \tag{6-5}$$

where j is the designation of different census tracts and n is the total number of census tracts with risk greater than 1 in 10 million.

3. Hazard Index

Chronic noncancer effects are evaluated by comparing exposure to RfDs. The RfDs, as discussed above, identify a lifetime exposure rate that would be without adverse effect, even for sensitive individuals. The measure of chronic effects is the hazard index (HI), which is the exposure from each pollutant divided by the RfD for that pollutant,

$$\text{HI} = \frac{\text{Exposure}}{\text{RfD}} \tag{6-6}$$

One complication that occurs for chronic effects is that exposure to multiple pollutants may need to be considered in cumulative effects. The HI is determined for each target organ or toxic endpoint. Target organs or effects include the liver, kidney, central nervous system (CNS), blood, etc. The determination of significance for chronic effects is the sum of the exposure divided by RfD for all pollutants with the same endpoint,

$$\text{HI} = \sum_{k} \frac{\text{Exposure}}{\text{RfD}_k} \tag{6-7}$$

where k refers to the chronic pollutants having the same target organ endpoint. An HI of less than 1 indicates that chronic effects are not significant for that target organ endpoint.

4. Acute Effects

Acute effects are evaluated by comparing maximum concentrations to acute standards. Acute standards are developed from short-term toxicity studies. They are typically limited to inhalation, since ingestion is primarily chronic. Acute environmental effects are rarely a concern from normal operations. This is because exposures for the public are usually much less than those for employees, and an acute health effect would consequently be observed for employees much sooner than the public.

For unintentional releases due to equipment failure or other catastrophic events, there may be acute impacts on the population. These impacts are not discussed here, but have been examined elsewhere.[11]

F. Uncertainty

As you can note from the above discussion, the calculation of a health risk associated with a facility or source requires substantial knowledge about several topics and numerous assumptions. With each of the pieces of information used to carry out the health risk assessment, and with each assumption, there is a level of uncertainty. None of the numbers used is known exactly, and in most instances, as noted above, effort is made to use a conservative estimate to assure that the public health is protected. However, the repeated use of conservative assumptions increases the level of conservatism after each calculation so that the final health risk estimate obtained can significantly overestimate the actual risk that may be experienced by an individual.

In an attempt to overcome this limitation on health risk assessment calculations, *Monte Carlo* analysis techniques have been proposed.[12] In the *Monte Carlo* analysis, a computer program is developed that calculates health risk using all of the assumptions and models identified above. The *Monte Carlo* process then calculates the health risk several thousand times. In each calculation, the value of each input parameter is varied randomly within its likely range. The results of these calculations are treated like data points in an experiment and an average and standard deviation is calculated. Using this information, the probability of the risk being a certain value can be calculated, and this information can be used to increase the quality of information available for risk management decisions.

G. Risk Management

Once the calculation of health risk has been completed, a decision must be made on whether the risk is significant. Usually this decision is made by the permitting agency. However, the health risk assessment may also be used by a decision-maker within a company to decide whether to go forward with a project.

Risk assessments are also used to manage or reduce the risks of a project. Often there are only a few sources within a facility and a few pollutants that contribute substantially to the health risk from a project. Consequently, the facility risk can often be reduced by addressing emissions from only a few sources, or only a few materials, feedstocks, or fuels. The culpable sources and pollutants are clearly identified in a health risk assessment, and efforts to reduce risk need only focus on these.

The determination of whether a calculated risk is significant is often established by the permitting agency in its regulations. State and federal regulators have made risk decisions at several levels historically.[13] The Food and Drug Administration has relied upon a one-in-a-million threshold for approval decisions for foods and drugs because of a Congressional mandate that has only recently been removed. Some decisions by EPA have approved risks as high as one-in-one-thousand. Most risk-based decisions are made with acceptable risks between one in ten thousand and one in a million, with an evolving regulatory threshold at one in one hundred thousand.

REFERENCES

1. Clean Air Act Amendments of 1990, Title III, Section 112(d)(2).
2. Ibid., Section 112(d)(3).
3. Miller, Susan, Radian Corporation, December 27, 1995.
4. Guidelines for MACT Determination Under Section 112(g), EPA-450/3-92-007b.
5. 59 *Federal Register* 15504, April 1, 1994.
6. *Air Pollution Consultant,* Vol. 4, July/August, 1994, p. 4.29.
7. U.S. EPA *Cancer Risk Assessment Guidelines.*
8. U.S. EPA *Superfund Risk Assessment Guidelines.*
9. California Air Pollution Control Officers (CAPCOA) Air Toxics "Hot Spots" Program Risk Assessment Guidelines, January, 1992.
10. U.S. EPA *Cancer Risk Assessment Guidelines.*
11. Hanna, S. R. and Drivas, P. J., *Guidelines for Use of Vapor Cloud Dispersion Models,* Center for Chemical Process Safety of the American Institute of Chemical Engineers, New York, 1987.
12. Alberts, Michael T. and Fred O. Weyman, "Characterizing Uncertainty in the Risk Assessment Process", in Winston Chow and Katherine K. Connor, *Managing Hazardous Air Pollutants, State of the Art,* Lewis Publishers, Boca Raton, FL, 1992, p. 319.
13. Travis, Curtis C., Edmund A. C. Crouch, Richard Wilson, and Ernest D. Klema, "Cancer Risk Management, A Review of 132 Federal Regulatory Decisions," *Environmental Science and Technology,* 21, 5, 1987.
14. 1990 Clean Air Act Amendments Title III, Section 112(g).

7

PERMITTING STRATEGY

I. INTRODUCTION

This chapter addresses how to develop a strategy or plan to obtain permits for a source of air emissions. The discussion includes all of the elements that may be necessary in a permitting strategy, even though not all elements are always required in a given case. All of the possible elements should be considered. Unnecessary elements can then be eliminated, rather than proceeding into the permitting activities and discovering later that an element omitted is suddenly critical.

The sections in the chapter chronologically examine the elements in a permitting strategy. First, goals in addition to obtaining a permit are discussed. The focus is on the other values that must be satisfied while a permit is being sought. The permitting team is discussed next, with an explanation of the role of each team member. Identifying permitting requirements was discussed in Chapter 1 for air permits, but most large projects require more than just air quality permits, and in this section these other requirements are identified. The permitting schedule is also described. It is developed by determining the time required to complete each element. Permitting costs are briefly discussed, along with the considerations in permitting that most affect costs. Project changes during permitting are addressed because they are a practical reality for a large project. If project changes are unexpected, both the costs and time required to obtain permits can increase quickly. Finally, this chapter discusses the process of negotiating permit conditions, and how this final step in permitting should be addressed.

SEVERAL QUESTIONS ARE IMPORTANT TO ANSWER IN DEVELOPING A PERMITTING STRATEGY

- What is your reputation with the permitting agency and with the public?
- How big a hurry are you in?
- What information do you have available?
- What is your budget?

Before beginning this discussion, it is important for the applicant to remember the orientation of the regulatory agency and for regulatory agency staff to understand the orientation of the applicant. Two sets of suggestions to remember in permitting a project are shown below. The first identifies suggestions for the applicant during the permitting process. The second identifies suggestions for the regulatory agency staff interacting with the applicant. These lists should be reviewed periodically in the course of the permitting process to keep perspective and a sense of reality during crises that may arise.

SUGGESTIONS FOR THE APPLICANT IN WORKING WITH THE PERMITTING AGENCY

1. Do not assume that you are popular.
2. Follow legal requirements.
3. Do not ask for special consideration or favors.
4. Be forthright and honest.
5. Do not offer gratuities.
6. Minimize political influence.
7. Avoid adversarial situations — seek consensus and cooperation.
8. Establish realistic project time-frames, goals, and objectives.
9. Provide accurate and thorough information.
10. Resolve issues at the lowest possible level.

II. PERMITTING GOALS

The primary goal of the permitting process is to obtain an air quality permit. At the same time, applicants will have other goals that also must be satisfied. These goals include minimizing costs, minimizing the time required to obtain the permit and build the project, limiting disclosure of confidential or competitive information, and limiting other risks of various sorts. Each of these is discussed below.

SUGGESTIONS FOR THE REGULATOR IN WORKING WITH A PERMIT APPLICANT

1. Assume that you are not popular.
2. Establish clear standards and requirements.
3. Treat all applicants equally.
4. Be forthright and honest.
5. Do not take gratuities, no matter how small. (Hint: There's no free lunch.)
6. Maintain a professional posture with elected officials.
7. Avoid compromising situations
8. Respond to calls, inquiries, and requests as soon as possible.
9. Use press releases when possible.
10. If interviewed, be prepared and avoid offensive or "off the record" comments.

A. Minimum Cost Goal

A common goal during permitting is to minimize costs. Costs involved can include the costs to obtain the permits, the capital cost of the project, and the operating and maintenance costs of the project. As discussed in Chapter 3, control equipment costs can be a significant fraction of total project capital and operating costs, and costs for emission reduction credits discussed in Chapter 4 are often comparable to control system costs. As the permitting strategy is developed, there may be choices about increasing expenditures during permitting. For example, hiring specialists to develop BACT analyses or obtain offsets may be necessary in order to reduce expenditures for control equipment or offset purchases. If minimizing costs is a goal, it will be important to look at the tradeoffs between later capital and operating expenses and upfront expenses associated with permitting. It is also important to have a clear understanding of the limits of expenditures. Project budgets must be flexible to accommodate unexpected but necessary costs. However, unconstrained permitting budgets can grow to be an unacceptably large fraction of project costs on the one hand, and an artificially constrained permitting budget may prevent actions that would increase permitting costs but ultimately limit or reduce the project costs. In the extreme case, increased permitting expenditures may enable a project to obtain a permit that would otherwise be impossible.

PERMITTING GOALS INCLUDE MORE THAN GETTING A PERMIT

- Minimum costs
- Minimum time
- Minimum risk
- Minimum disclosure

If permitting costs are minimized, it will usually take longer to obtain a permit. Initial permit preparation costs can be minimized by preparing a "minimalist" or "bare bones" application. This leaves the agency with the burden of identifying additional information needed and requesting it from the applicant. This "bare bones" approach usually takes longer because of the exchange of communication between the agency and the applicant. However, it circumvents the problem of presenting extraneous information to the agency that may take extra time to review and may not be directly relevant to the permit to be issued. The "bare bones" approach can also leave the impression with the agency that the applicant is not acting in good faith.

A variant on the "bare bones" approach is to prepare one permitting document that describes the entire project and can be used by all of the agencies for all of the permits required. Although the document is longer and has more information needed by any one agency, it is presumed that money can be saved because several subject-specific documents will then be unnecessary. My experience has been that this approach results in additional time required for permitting and additional conditions in the permit. The agency staff require more time to review the entire document because it takes longer to find information needed, and they feel obliged to read the entire document to be sure that information relevant for their analysis is not contained in a section that would appear to be irrelevant. Additional questions may come from the agency regarding the extra sections of text, and additional conditions may be proposed by the agency to assure that portions of the project separate from air quality issues do not result in air quality impacts.

Another approach to minimizing permitting costs is to avoid or completely refuse to use consultants to assist in application preparation. If the application is simple and the applicant's relationship with the regulatory agency is good, this approach can result in very low cost. However, if the project is complicated, it may take considerable time for staff in the applicant's organization to learn the techniques needed to complete an adequate application, and overall costs will be less if a qualified consultant is involved.

One area in which substantial upfront cost is justified is in the selection of control technologies. As discussed in Chapter 3, BACT is required for most new sources that will be permitted in the 1990s. The permitting agency will look carefully at the selection of control technology to be sure that it constitutes BACT and will expect comprehensive justification of the technology selected, especially if it does not appear to provide the greatest level of control. A thorough preliminary design that considers a broad range of alternative control technologies is justified, since subsequent rationalization of the technology selected will need to be provided to the permitting agency. Failure of the applicant to look at all options could result in a mandate, late in the permitting process, for selection of a control technology different than originally proposed. This would consequently result in requirements for design revisions at a point when bid specifications are usually being prepared.

B. Minimum Time

A second common permitting goal is to minimize the time required to obtain a permit. In almost every instance, this goal is inconsistent with a goal of minimizing the cost of permitting. Minimizing permit acquisition time requires extraordinary effort in preparation of the supporting analyses and documentation for permitting. Also, using a consultant is almost always required unless the applicant's staff can be pulled away from other responsibilities for a substantial portion of their time to address the application requirements. To ensure accuracy and completeness in the submitted application as well as subsequent information provided to the permitting agency, extra effort and increased costs are required.

An additional cost burden associated with a minimum time goal stems from permit conditions. An agency aware of urgency on the part of the applicant may require extraordinary permit conditions. These conditions can be agreed to in order to minimize the time required in negotiation, or they can be resisted, with additional time required to finally reach a compromise acceptable to both the applicant and the agency.

An approach that is sometimes considered by an applicant to achieve a minimum time for permitting is to utilize political influence to speed up the process. This approach must be used with considerable caution, since most jurisdictions have explicit requirements that must be met for a permit to be approved, and circumventing any of those requirements through political influence may result in subsequent challenges to the permit in the courts. This leads to the next goal, minimum risk.

C. Minimizing Risk

The goal of minimum risk in permitting is accomplished by taking steps to assure that the risk of legal challenge to the permit is minimized. This approach is most often used when there is opposition to a project (see the discussion below on Community Relations). This is sometimes referred to as an *abundantly cautious* approach. In providing documentation, assessing environmental impacts, or providing opportunities for public participation, everything conceivable is done to assure that any challenge to the permit in court will be overcome. The minimum risk goal is often adopted as secondary to primary goals of minimum cost or time, but it is seldom compatible with either of those goals. Additional documentation, assessment, or participation all require extra time and money, and thus compromise those other goals. On the other hand, if an issued permit is challenged in court, the minimum cost or time goals can be significantly impacted by legal fees and time in litigation. Therefore, it is necessary for the applicant to look carefully at both short-term and strategic considerations when evaluating minimum risk against other goals.

D. Minimizing Disclosure

Another permitting goal that may be considered is to limit the disclosure of information about the facility or process for which emissions are being permitted to the greatest extent possible. This goal is most common when the process is new, constitutes a trade secret, and/or may be a strategic opportunity for a business. Both federal and state laws require that the emissions from a process be quantified. Newer rules about hazardous air pollutants (HAPs) may impose quantification requirements for not only criteria pollutant emissions but also for HAPs (see Chapter 6). Some processes emit small quantities of input materials that are essential to the process and are part of the trade secret for that process, such as some emissions from semiconductor fabrication processes.

INCLUDING TRADE SECRETS IN AN APPLICATION SHOULD BE AVOIDED WHENEVER POSSIBLE

- Emissions data cannot be protected as a trade secret.
- Process information can be protected, but safeguards are weak.

Another motivation for a minimum disclosure goal would be a process that is new or innovative but after disclosure could be replicated easily by competitors. This is probably the most difficult minimum disclosure situation, and often the applicant must be satisfied with the advantage that a competitor must also obtain a permit, and thus is that far behind. Limitations on disclosure can be imposed by declaring portions of the application, such as operating parameters during source testing, trade secrets. Regulatory agencies have provisions for the maintenance of confidential business information that is submitted as part of an application. These provisions should be reviewed by counsel to be sure that any special marking or packaging of confidential material meets agency requirements, and that counsel is prepared to resist any requests under the Freedom of Information Act or similar state freedom of information statutes.

Another approach may be to strategize for minimum time of public disclosure about the application to limit opportunities for others obtaining information about a proposed project. Applications can be prepared without disclosure, with any questions for regulatory agency staff presented by a consultant or outside counsel, and submission of the application can be the first public action.

A minimum disclosure goal may be consistent with one of the other goals (minimum cost, time, or risk) in most circumstances, but it presents an additional constraint on the permitting team.

III. PERMITTING TEAM

A team of people is required to obtain a permit for an air emission source. To obtain a permit for a single piece of equipment in a preexisting facility, the team may consist of two or three persons who only work on the permit application part time. But for a new facility with large emission rates (such as a new electric power plant) or numerous sources (such as a semiconductor manufacturing facility), the permitting team will have several participants, and one or more of them may be committed full time for an extended period during the application process. The membership of a large permitting team is shown below.

PERMITTING TEAM MEMBERS

- Project manager
- Environmental counsel
- Technical specialists
   Emissions
   Control technology/BACT
   Air quality impact analysis
   Environmental impact analysis
   Offsets
   Air toxics/health risk assessment
- Community relations specialist

A. Project Manager

The single most important person on a permitting team is the project manager for the applicant. To obtain the permit and meet the permitting goal or goals discussed above, the project manager needs to have the authority to negotiate permit conditions for the applicant. Often a project manager has a limit to which he or she can commit the company, and for major decisions, such as major control technology commitments, will need to obtain the approval of a more senior manager. However, the most expeditious permitting — with either minimum time or minimum cost goals — is achieved when the project manager has a very high level of authority and can negotiate almost all of the conditions of the permit without consultation.

A permit can certainly be obtained without vesting all authority in a single individual, and in many permitting efforts it is not possible to grant extensive authority to one person. Examples of those situations would be joint ventures in which project ownership is shared by two or more partners, and approval by those partners would be necessary for major cost decisions. Another similar situation would be a partnership in which cost minimization was a goal because of limitations on available capital. This is often the case with small cogeneration projects. In those circumstances, every effort should be made to grant the maximum discretion to the project manager within the financial constraints of the project owners.

B. Environmental Counsel

To fulfill permit requirements, it is necessary that some member of the permitting team be thoroughly familiar with the regulatory requirements. For a small application effort, this knowledge may be held by the project manager because of his or her previous experience with the permitting agency. But in most major permitting projects, environmental counsel fulfills this role. Counsel is needed to provide an accurate interpretation of the regulatory and other legal requirements that must be satisfied for a particular project to be permitted. Counsel should have a good understanding of the rules in the jurisdiction in which the permit is being sought as well as the personalities of the decision makers who will ultimately approve the permit. As discussed below, any major project needs to obtain other permits in addition to the air permit, and counsel should also have the knowledge to recognize other requirements that must be met in concert. In addition, presentations by the applicant at public hearings are usually orchestrated by counsel based upon his or her knowledge of the process that is occurring. Environmental counsel can be in-house within the applicant organization or can be outside counsel. There is not a general preference for either; however except in companies frequently involved in permitting, in-house counsel usually does not have the level of environmental experience that can be found in outside counsel who specialize in environmental work.

C. Technical Specialists

For a major permitting project, technical specialists are needed for each of the areas discussed in the previous chapters of this book. In a small project, the project manager may know enough about the process to estimate emissions, and the emissions may be too small to require performing an air quality impact analysis (Chapter 5) or obtaining offsets (Chapter 4). However, large projects will benefit from specialists who can address the several elements needed in a successful permit application. As with environmental counsel, these experts may be partially or entirely drawn from staff in the applicant company, or they may be consultants. The emission estimates (Chapter 2) and control technology determination (Chapter 3) are most often the elements that are retained in-house because these areas benefit from staff familiarity with the processes involved. However, even here, consultants may be used for peer review and quality assurance. Air quality impact analyses (Chapter 5) and hazardous air pollutant or other air toxics analyses (Chapter 6) are most often done by consultants. In a major permitting project, there will be experts in the process and control technologies from within the company, in an engineering design organization, and in a consulting organization all of whom are participants on the permitting team.

D. Community Relations

As permitting requirements have increased the opportunity for public involvement (Title V permits must be noticed to the public for 30 days before they are issued),[1] it is increasingly important that the applicant make provisions for dealing with the public who are participants in permit decision making. Often in a small permitting project, the project manager accepts responsibility for dealing with the public. Even in a large project, it is best that the project manager be the spokesperson for the project. When dealing with the public, it is often useful to have support from a community relations expert. The applicant's positive reputation with the public in the vicinity of the proposed facility or project is critical to ensuring successful permitting of a controversial project. Techniques such as community opinion surveys, newsletters, facility sponsorship of and participation in community activities, preparation of press releases, and notifying public officials can all be developed or coordinated by a community relations professional.

One of the "public" groups that can be very important in a permitting project are the employees within the company. If the company is a large employer in the community, employees can be influential in community opinion. Employee newsletters and meetings with employees can reduce anxiety associated with a new project and provide employees with information to enable them to become advocates for the project within and outside the plant.

E. Team Communications

For a permitting team to be effective, it must have good communications as well as the competent membership described above. Often, when outside consultants or counsel are utilized as part of the permitting team, an applicant will only provide to these "outsiders" the information the applicant perceives is necessary to carry out the specific assignment. However, in permitting a complicated project, wide-ranging considerations can affect a single permit application. Because of this, comprehensive sharing of information is essential to ensuring that permitting is successful. Some of the methods used for ensuring adequate communication are to have regular meetings of the team, teleconferencing, and weekly newsletters. The techniques that have been developed elsewhere for team building are effective with permitting teams as well. Even when the agenda is short, regular meetings result in useful information being exchanged and often creative interchanges to solve problems that arise.

Another requirement for a strong permitting team is that there be clear authority held by the project manager. As in other managerial situations, it is vital that this manager continuously be open to receiving new data. At the same time, the project manager must make the decisions.

A third requirement for the permitting team is that it be open to information from outside sources. If a project is controversial, the project team must be aware of the arguments being raised by opponents and be clear about either developing appropriate rebuttal or being willing to alter the project to address legitimate concerns.

IV. NON-AIR QUALITY PERMITTING REQUIREMENTS

Early in the development of a permitting strategy, the permits or other approvals that are necessary before the project can be completed must be identified. For a small project, only the air permit and perhaps a building permit may be necessary. For a large project, there may be a dozen approvals necessary from an equal number of agencies. Table 7-1 shows a list of potential approvals that may be needed for a project, and the federal statutes governing those approvals. In each state, there may be additional or supplemental approvals in each of these subject areas, but in all cases the federal requirements are the minimum requirements. State and local jurisdictions cannot exempt an applicant from those requirements, but those jurisdictions can add to the requirements either in stringency or breadth. Each of these areas is discussed below.

A. Aesthetics

Most federal legislation recognizes aesthetics as important, but the language in most acts is general. The Federal Land Policy and Management Act of 1976 has scenic values included in Section 8, and the Wild and Scenic Rivers Act of 1968[2] has a primary emphasis on retaining the scenic value of rivers. Two guidelines published by USDA[3] and BLM[4] are frequently used.

B. Biological Resources

Currently major controversies are being played out over the Endangered Species Act of 1973.[5] The act provides for the identification of endangered and

Table 7-1   Environmental Statutes That Could Affect Air Permitting Projects

Impact area	Governing law or guideline	Description	Legal citation
Aesthetics	Federal Land Policy and Management Act of 1976, Section 8	Most federal legislation recognizes aesthetics as important, but the language in these acts is general.	
	Wild and Scenic Rivers Act of 1968	Primary emphasis is on retaining the scenic value of rivers.	PL 90-542, 82 Stat 906, 16 USC 1271
Biological resources	Endangered Species Act of 1973	Provides for the identification of endangered and threatened species, and for the protection of those species.	PL 93-205, 87 Stat 884, 16 USC 1531. Regulations in 50 CFR 17.1.
Cultural resources	National Historic Preservation Act of 1966.	See especially Section 106 establishing the National Register of Historic Places.	PL 89-665, 80 Stat 915, 16 USC 470 (amended in 1976 by PL 94-222 and 94-458). Regulations in 36 CFR 800
	American Indian Religious Freedom Act of 1978		43 USC annotated section 1996
	Native American Graves Protection and Repatriation Act of 1990		PL 101-601
Environmental impact analysis	National Environmental Policy Act of 1970	Requires preparation of an environmental impact statement for any "federal action which could significantly affect the human environment" (Sec. 102).	PL 91-190, 42 USC 4321
Hazardous materials	Superfund Amendments and Reauthorization Act of 1986 (SARA)	Title III established an emergency planning and response program that requires businesses to report storage, handling or production of significant quantities of hazardous or acutely toxic substances.	PL 990-499, 42 USC Section 11001 et seq. Regulations are in 29 CFR 1910 et seq.
Hazardous wastes	Resource Conservation and Recovery Act (RCRA) of 1976 and RCRA Amendments of 1984	RCRA addresses both hazardous and solid wastes. This legislation governs the treatment, storage, and disposal (TSD) of hazardous wastes, and requires permits for TSD facilities.	PL 94-5809, 90 Stat 95, 42 USC 6901. Regulations are in 40 CFR 260 et seq.

Category	Act	Description	Citation
Land use	Comprehensive Environmental Response, Compensation, and Liability Act of 1980 (CERCLA)	CERCLA addresses uncontrolled hazardous waste sites. It is seldom an issue in a newly built facility, although CERCLA could be involved if the site contains wastes from previous activities.	PL 96-510, 94 Stat 2767, 42 USC 9601. Regulations are in 40 CFR Parts 300 to 305.
	Federal Land Policy and Management Act of 1976	Covers lands administered by the Bureau of Land Management and the Forest Service.	
	Coastal Zone Management Act of 1972	Regulates land use in coastal areas, usually through state coastal zone management boards.	PL 92-583, 16 USC 1451
Noise	Noise Pollution and Abatement Act of 1970	Identifies noise pollution as a problem, but does not impose a control program.	Passed as Title IV of the Clean Air Act (PL 91-604, 42 USC 1857)
	Noise Control Act of 1972	Assigned primary noise control responsibility to local government.	PL 92-842, 42 USC 4901
Paleontological resources	Antiquities Act of 1906	Protects fossils from former geologic periods.	
Risk of upset	Superfund Amendments and Reauthorization Act of 1986 (SARA) Title III	Establishes an emergency planning response program.	42 USC 11001
	Clean Air Act Amendments of 1990	Title III (Toxics) includes a section on managing the risk of upsets.	PL 101-549, 42 USC 7401 et seq., 40 CFR Part 68
Solid wastes	Resource Conservation and Recovery Act of 1976 (RCRA), Title II, Subtitle D	Establishes the structure for control of solid waste disposal facilities.	PL 94-5809, 90 Stat 95, 42 USC 6901. Regulations are in 40 CFR 260 et seq.
	Pollution Prevention Act of 1990	Requires that goals be established for recycling at a facility.	PL 101-508 (Omnibus Budget Reconciliation Act of 1990)
Traffic and transportation	Hazardous Materials Transportation Act of 1974	Required that criteria be established for hazardous material transportation.	49 CFR Subtitle B. Chapter I, subchapter C and Chapter III, subchapter B.

Table 7-1 Environmental Statutes That Could Affect Air Permitting Projects (Continued)

Impact area	Governing law or guideline	Description	Legal citation
Water quality	Federal Water Pollution Control Act of 1972	Established the National Pollutant Discharge Elimination System (NPDES) permitting for discharges into receiving waters.	PL 92-500, 33 USC 1251, 86 Stat 816
	Clean Water Act of 1977	Revised and tightened FWPCA and addressed quality standards for receiving waters.	PL 95-217, 91 Stat 1567, amended in 1987, PL 100-4, 101 Stat 76
	Safe Drinking Water Act of 1974	Regulates groundwater and drinking water supplies.	PL 93-523, 42 USC 300, 88 Stat 1661. Regulations are in 40 CFR 122, 123, 146, and 147.
	Rivers and Harbors Act of 1899	Section 404 requires a Corps of Engineers permit for alteration of navigable waters.	33 USC 401-413
Worker health and safety	Occupational Safety and Health Act of 1970	Established worker health and safety requirements.	PL 91-596, 84 Stat 1590. OSHA regulations start at 29 CRF 1900

Source: Radian International LLC

PERMITTING USUALLY INCLUDES MORE THAN JUST MEETING AIR QUALITY REQUIREMENTS

- Aesthetics
- Biological resources
- Cultural resources
- Environmental impact analysis
- Hazardous materials
- Hazardous wastes
- Land use

- Noise
- Paleontological resources
- Risk of upset
- Solid waste
- Traffic and transportation
- Water quality
- Worker health and safety

---

threatened animal and plant species, and for protection of those species. Several revisions to this legislation have been proposed.

C. Cultural Resources

Several statutes have been enacted to protect cultural resources. They include the National Historic Preservation Act of 1966,[6] especially Section 106 establishing the National Register of Historic Places and Executive Order 11593, Protection of the Cultural Environment.[7] The American Indian Religious Freedom Act of 1978[8] and the Native American Graves Protection and Repatriation Act of 1990[9] also address cultural resources.

D. Environmental Impact Analysis

The National Environmental Policy Act of 1970[10] includes a requirement for the preparation of an Environmental Impact Statement (EIS) for any "federal action which could significantly affect the quality of the human environment."[11] An EIS must address all of the impacts of a proposed project that could be anticipated, including those listed here as well as socioeconomics, water availability, public safety, and other areas. Many of the states have passed "little NEPAs" that impose similar environmental impact assessment requirements on actions taken at the state level.

The determination of projects for which an EIS must be prepared has narrowed in the last decade because of exclusions included in legislation[12] or a delegation of approval to the state level so that no federal action is involved.

E. Hazardous Materials

The Emergency Planning and Community Right-to-Know Act (EPCRA) was passed as Title III of the Superfund Amendments and Reauthorization Act of 1986 (SARA).[13] Title III established an emergency planning and response program that requires businesses to report storage, handling, or production of significant quantities of hazardous or acutely hazardous substances. A summary of the provisions in EPCRA is shown in Table 7-2.

Table 7-2 Summary of EPCRA Requirements

Statutory Section	301-303	304	311	312	313
Topic and regulatory citation	Emergency Planning (40 CFR 355.30)	Emergency Release Notification (40 CFR 355.40)	Hazardous Chemical Inventory (40 CFR 370-21)	Hazardous Chemical Inventory (40 CFR 370.25)	Toxic Chemical Release Reporting (40 CFR 372)
Chemicals covered	EHSs (40 CFR 355, App. A&B)	EHSs and CERCLA hazardous substances (40 CFR 355 and 302.4)	Hazardous chemicals (no specific list)	OSHA hazardous chemicals (no specific list)	Toxic chemicals and chemical categories (40 CFR 372.65)
Thresholds	TPQ (chemical-specific)(1 to 10,000 lbs)	Reportable quantities (chemical-specific)(1 to 5,000 lbs)	10,000 lbs; if EHS, 500 lbs or the TPQ, whichever is less	10,000 lbs; if EHS, 500 lbs or the TPQ, whichever is less	25,000 lbs manufactured or processed; 10,000 lbs otherwise used
Reporting	Notification letter to SERC and LEPC and additional planning information	Verbal and written notification to SERC(s) and LEPC(s)	MSDSs or list of chemicals to SERC, LEPC, and LFD	Tier I, Tier II, or state-equivalent form to SERC, LEPC, and LFD	Form R report to U.S. EPA and SERC
Deadlines	3 March 1994 (notification) and 3 August 1994 (additional information)	Immediately following the release event (beginning on 1 January 1994)	3 August 1994	On or before 1 March 1995 and annually thereafter	On or before 1 July 1995 and annually thereafter
De minimis Exemption	≤1%	None	≤1% or ≤0.1%	≤1% or ≤0.1%	≤1% or ≤0.1%

CERCLA = Comprehensive Environmental Response, Compensation, and Liability Act
CFR = Code of Federal Regulations
EHS = extremely hazardous substance
LEPC = Local Emergency Planning Committee
LFD = local fire department
MSDS = material safety data sheet
OSHA = Occupational Safety and Health Administration
SERC = State Emergency Response Commission
TPQ = threshold planning quantity
U.S.EPA = United States Environmental Protection Agency
Source: U.S. EPA EPCRA Training Manual, n.d.

F. Hazardous Wastes

Federal requirements for management of hazardous wastes was established in the Resource Conservation and Recovery Act (RCRA) of 1976[14] and the RCRA Amendments of 1984.[15] RCRA permitting is often required for a new facility, but is usually administered by a state agency that has received delegation of the federal authority. Uncontrolled hazardous waste sites are addressed in the Comprehensive Environmental Response, Compensation, and Liability Act of 1980 (CERCLA).[16] CERCLA is seldom an issue in a newly built facility unless the construction site has contamination from previous use. When this is the case, CERCLA issues can dominate the permitting process.

G. Land Use

Although there are several land use laws and policies that have been established at the federal level, most land use approvals are at a regional or local level through use permits and zoning requirements. The Federal Land Policy and Management Act of 1976 covers lands administered by the U.S. Bureau of Land Management and the U.S. Forest Service. If a project is in a coastal area, it may be subject to the Coastal Zone Management Act of 1972[17] and a local or regional coastal zone management agency.

H. Noise

Federal noise guidelines have been issued by the Environmental Protection Agency, which identified a level of 55 decibels as adequate to protect outdoor activities against interference due to noise. The Noise Pollution and Abatement Act of 1970 was passed as Title IV of the Clean Air Act of 1970.[18] The Noise Control Act of 1972[19] assigned primary noise control responsibility to local government. Many local jurisdictions have established noise regulations or enforce noise standards through nuisance provisions of health and safety ordinances.

I. Paleontological Resources

Fossils from former geologic periods are protected under the Antiquities Act of 1906. The U.S. Bureau of Land Management issued a 1978 memorandum outlining significance criteria for paleontological resources, and the Society for Vertebrate Paleontology distributed a draft set of guidelines in 1989 that identified acceptable professional practices in the conduct of paleontological resource surveys, data recovery, analysis, and curation. In areas where paleontological resources are suspected because of previous discoveries, many projects retain a paleontologist who can be quickly brought to a site to evaluate possible discoveries during construction.

J. Risk of Upset

The hazardous materials management programs under EPCRA include provisions for emergency planning and response. There are also requirements for managing the risk of upset that could result in release of acutely hazardous materials in Title III of the Clean Air Act Amendments of 1990.[20] Article 80 of the Uniform Fire Code includes provisions for storage and handling of hazardous materials. For it to apply, it must be adopted by reference by the local fire district.

K. Solid Wastes

Solid waste disposal facilities are regulated under Subtitle D of RCRA. As noted above, RCRA has usually been delegated to state governments. Management of solid wastes at a facility is addressed in the Pollution Prevention Act of 1990.[21] This Act requires that goals be established for recycling solid wastes at a facility.

L. Traffic and Transportation

The Hazardous Materials Transportation Act of 1974 directed the Department of Transportation to establish criteria and regulations for the safe transportation of hazardous materials.[22] Traffic and transportation is usually addressed as part of an EIS as well.

M. Water Quality

The Federal Water Pollution Control Act of 1972[23] and the Clean Water Act of 1977,[24] along with amendments in 1987,[25] address water quality for receiving waters of the United States The FWPCA also resulted in the establishment of the National Pollutant Discharge Elimination System (NPDES) permitting for discharges into receiving waters. The Safe Drinking Water Act of 1974[26] regulates groundwater and drinking water supplies. The Rivers and Harbors Act of 1899[27] includes the well-known Section 404, which requires a permit from the Corps of Engineers for obstruction or alteration of navigable waters of the United States. Navigable waters include almost any stream, and the requirement for a 404 permit is often a federal action that triggers an EIS. Section 404 permitting has not been delegated to the states.

N. Worker Health and Safety

The most comprehensive worker health and safety requirements are under the Occupational Safety and Health Act of 1970.[28] Usually, occupational safety and health is addressed independently from environmental issues, and many environmental regulations (air quality regulations in particular) are only applicable outside the property line of the facility. Nonetheless, occupational standards can be a source of numerous potential violations even for a well-run facility. OSHA criteria have sometimes been used as a basis for ambient standards for toxic pollutants.

V. PERMIT CONTENT

The contents of a permit application will be different for each jurisdiction and for each permit type discussed in Chapter 1. Typical contents for an application are shown in the box below. The regulatory agency usually provides an application form that should be completed, but for a major source or major modification, additional information will be necessary to provide the permitting agency with enough information to make a decision.

ELEMENTS OF AN AIR PERMIT APPLICATION

- Project description
- Emissions estimates
- Applicable regulatory requirements
- BACT analysis
- Air quality impact analysis
- Offsets and mitigation measures
- Toxic air contaminant risk screening
- Air quality related values analysis

Not all of these elements are required in every application.
Only required elements should be included.

If there are questions on the application form that are unclear, a call to the agency is worthwhile. Usually permitting staff have been asked similar questions before, and they are generally very willing to clarify the application form so that the information submitted is clear and complete. Some agencies have prepared very complete instructions on what should be submitted in the permit application for particular source types.[29]

The application should include a description of the project being proposed. The complexity of the description should be consistent with the complexity of the project. However, in every case, it needs to include an identification of each pollutant-emitting unit or process, with an estimate of emissions. Usually the details of the emission estimate are included as an appendix.

The applicant must be aware that the emission estimate provided will probably be treated as an emission limit in the permit that is issued. Consequently, this estimate needs to be of the maximum emissions anticipated. The disadvantage of estimating emissions too high is that a high estimate may trigger additional requirements, such as offsets or BACT, and will usually result in higher annual fees that are based on the emission level. The emission estimate must be at a level that allows the operations anticipated at the plant, since operation at emissions above the permit limit constitute a violation of the permit.

A list of applicable regulatory requirements is a required part of Title V permit applications, but it is worthwhile to include in any permit application to identify for the agency staff which rules you consider applicable to your facility. The agency may disagree with your interpretation of the rules. However, having your interpretation in front of the reviewer can guide their thinking and speed up the review process. It will also allow more clarity in any disagreements between the applicant and the agency, and hopefully result in smoother and faster resolution.

If the source is subject to control technology requirements such as BACT or LAER, the selection of control technology analysis needs to be part of the application. Often the agency will not require a comprehensive top-down BACT analysis, but will accept a more abbreviated analysis. The level of the BACT analysis can be determined by reviewing applications that have recently been accepted by the agency. Presenting a more complicated analysis than is traditional for a particular permitting agency can increase the time required for review of the application, and may introduce information that results in questions from the agency that require additional time and work to answer.

If offsets are required for a permit, information about those offsets must be provided in the application. If the offsets are from within your plant, as a result of shutdown of an existing emission unit or a decrease in emissions, a complete description can be provided. However, if it will be necessary to purchase offsets, several additional considerations associated with purchase negotiations can become important, and disclosure of the anticipated source of offsets may not be prudent.

It is almost always necessary to pay a fee with the application. Usually the permitting agency will not consider submissions to be an application unless they are accompanied by the required permit application fee.

Decisions about how much detail to provide in the application should be based upon the permitting strategy and goals identified by the applicant. However, clarity should be sought in the application. Agency reviewers will be able to evaluate an application much more quickly if it is clearly presented.

VI. PERMITTING SCHEDULE

The permitting schedule is very important for project completion, because the air quality permit as well as most other permits must be obtained prior to the start of construction of a project.

Developing the permitting schedule requires that the permitting team determine the following:

- Permits required (both air quality and other environmental permits)
- What information is needed to complete permit applications
- How long it will take to prepare needed information
- Time required for review and approval of each permit
- Whether any permits or approvals must await a prior approval (often an environmental impact analysis must be completed before an air quality permit will be issued)

The permits and other environmental approvals and analyses required were discussed in the previous section. In most jurisdictions, the time allowed for an agency to review the application is specified, and if not, the agency will usually provide an estimate of the time required for review. Generally these estimates are reliable, but it may be helpful to write to the agency to confirm the time estimate provided so that there is a record of the estimate. It is helpful to have had experience with the agencies who are reviewing a permit to know whether review periods usually meet these time estimates. Unfortunately, there is little leverage that can be applied if an agency does not meet its mandated review times, since the agency has the discretion to deny a permit within the mandated time period if it is not finished with its review. Although no agency wants to take longer than the mandated time, the applicant has few options available but to agree to a longer review.

A meeting with the permitting agency early in the process is usually valuable. The meeting can be an opportunity to make the agency aware of when a permit application should be expected, and gives an opportunity to ask specifically what is expected in the application, what the length or level of detail of various sections should be, and to whom questions about the process should be addressed. Although an early meeting with the agency can be very beneficial in focusing the preparation of the air permit application, the meeting will likely be considered public, and can be listed in a summary of agency activities. If confidentiality is one of your permitting goals, such a meeting may not be appropriate for this reason.

The sequencing of permits will depend upon the regulations and precedent for such permits in the past. If there is an environmental impact statement or other such comprehensive environmental review, the environmental review process must be completed before other applications can be reviewed, in some cases, or in other cases before a permit can be issued. The intent of the comprehensive environmental review is that it provide input to the decision making process, and thus it needs to be available to permit writers so that they are aware of the breadth of impacts that are anticipated from the proposed project.

Set realistic times for preparation of the permit applications. This is especially important if the applicant has a permitting goal of a minimum time to obtain the permits. Permit applications can be submitted before final design is complete, and because of BACT requirements that may change the design, submitting the application before the final design is complete is usually preferred. Most experienced consultants can provide accurate estimates of the time required to prepare a permit application and will commit to those estimates provided the information needed from the project design is available.

With the above information, a schedule can be developed for permitting. MacProject[®30] provides an excellent graphic display of the sequencing of a project (Figure 7-1), as well as other scheduling aids. Other software provides similar capabilities and may be more familiar to the reader.[31]

Good scheduling software will indicate the critical path for the project. The sequence of permitting activities determines the critical path for permitting. Permitting management needs to focus on the activities and approvals on this path in order to minimize the time required for project approval. However, it is also important to recognize that other approvals may require nearly as much time

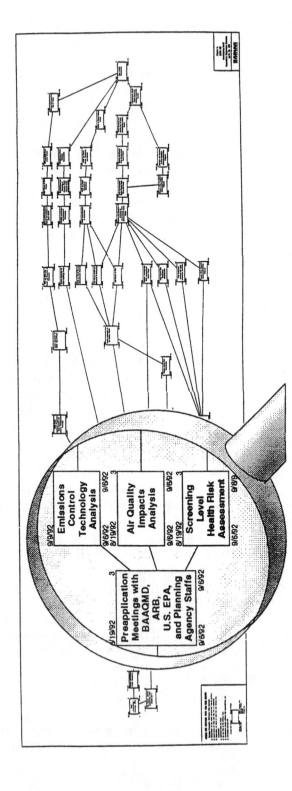

Figure 7-1 Several software packages will provide a graphic display of the permitting schedule. (Source: Radian International LLC.)

as the critical path. If this is the case, an unanticipated delay in some activity not on the critical path may result in it becoming critical.

Another important element in permit scheduling is to recognize that there may be other approvals needed for a project that are not environmentally related and that may determine the critical path for the project. For example, a power purchase agreement, fuel contract, or facility financing approval may be independent of environmental permitting and govern the overall schedule. However, more and more often, financing and other project approvals include consideration of environmental permitting. When this is the case, the development of an accurate environmental permitting schedule can be helpful in planning the rest of the project.

VII. PERMITTING COSTS

The cost to obtain environmental permits can range from a few hundred dollars for a permit application fee with no preparation costs beyond normal staff time, to several hundred thousand dollars in application fees, over $1 million in application preparation, and over $10 million in offset purchases. Because potential costs can be high, both advance planning and a substantial contingency are important.

PERMITTING COSTS CAN INCLUDE SEVERAL ITEMS

- Permit application fees
- Permitting team costs
- Mitigation measures and offsets

Permit fees are specified in regulations. They can be based on a variety of criteria. Usually there is a specified fee that must accompany the application. There are additional charges that are based on the size of the equipment or the level of emissions. Many agencies also charge for the time required to review an application, with typical rates ranging from $50 to $80 per hour. Charges based on review time can range from several thousand dollars to well over $100,000 for a very large project. Since agency staff are sometimes not accustomed to accounting for their time spent on particular projects, careful review of agency charges is usually worthwhile. Obtaining detailed documentation of hours charged may be difficult, and requesting such documentation in advance may improve the quality of records obtained.

Use of a consultant to prepare a complex application is usually more cost effective than hiring internal staff because of the indirect costs associated with hiring new or temporary staff. A consultant who has experience with the agency or agencies involved knows how the applications must be prepared and what information must accompany the application for it to be considered complete. A consultant also often has expertise, such as in air quality modeling, that is not needed very often for in-house activities and so is seldom among the capabilities of in-house staff. For permitting work, consultants usually work on a cost reimbursable basis. Fixed-price contracts may be negotiated for specific work elements in permitting, but usually the uncertainties in what the agency will require are too great to cover the risk of fixed price work without a very large contingency being requested by the consultant. Environmental consultants cost in the range of $10,000 to $20,000 per person-month, depending upon the expertise of the firm and the individuals involved.

One issue that arises in preparation of a permit application is what level of detail is needed to make the application complete. This becomes a cost issue because the greater the level of detail, the greater the cost of preparation. Determining the appropriate level of detail depends upon the behavior of the regulatory agency as well as the applicant's permitting goals. An effort to minimize cost of preparation by limiting the amount of information provided can save money in the short term, but may result in a longer time to process the application, since the agency will probably request additional information if they perceive that not enough information has been provided.

CHANGES IN THE PROJECT DURING PERMITTING WILL INCREASE COSTS, DELAY REVIEW, AND MAY COMPROMISE PERMITTING SUCCESS

- Requires redoing modeling
- New health risk assessment
- Additional offsets or other mitigation
- Reduced credibility with the agency and the public

VIII. PROJECT CHANGES DURING PERMITTING

There may be a number of reasons a project will change during the period between the start of preparation of the permit application and the issuance of a permit by the regulatory agency. If a change occurs before the application is submitted, the only impacts will be upon the cost of preparation of the application and the time required to make necessary modifications. However, if the change occurs after the application has been submitted, greater problems can result. When such a change occurs, it is important that the application be modified in a fashion that makes the changes clear, and that allows the application reviewer to utilize as much of the work previously completed to be retained as possible. This is not just an issue of saving costs billed by the agency for review. It also affects the agency, and individual staff members' attitude toward the applicant. The message that is conveyed to the agency when a change is made in a project application is that the applicant is uncertain, poorly organized, and disrespectful of agency staff time. These are not messages that will improve the applicant's relationship with the agency, nor will they tend to speed up the effort by the agency to complete the approval process. The more times a project is changed during permitting, the greater the scrutiny by the regulators, and the more questions that are likely to be asked about the details of the project, details that may be trade secrets or that could otherwise compromise the applicant's competitive position.

If minor changes in a project are anticipated, it may be better to await approval of the project before such changes are finalized. Often changes are needed in a project when it is under construction that were not anticipated during design. Agencies are sympathetic to the need for such changes provided emissions are not increased significantly and additional environmental impacts do not occur. However, if a change in the project was decided upon prior to permit approval, and the agency was not notified, serious consequences, including permit revocation and/or criminal charges may result.

PERMIT CONDITIONS ARE NEGOTIATED WITH THE AGENCY

- Draft conditions are developed by agency based upon application
- Ask agency for draft conditions
- Review conditions for enforceability and operating flexibility
- Conditions can be officially reviewed during the public comment permit, but conditions should be reviewed earlier

---

IX. PERMIT CONDITIONS

The final step in obtaining an air quality permit is the specification of permit conditions. Under the provisions of Title VII of the Clean Air Act Amendments of 1990, permits issued under the Title V federal permit program must include all of the regulatory requirements that must be met by the permittee for the duration of the permit (up to 5 years). Even before this program is implemented, air permitting agencies may, and usually do, include conditions in the permit that must be satisfied for the permit to remain in force.

The applicant should expect that all estimates of emissions and control efficiencies presented in the application will appear as permit conditions. There will probably be other conditions that could include constraints on startups and shutdowns, requirements for emission monitoring, and specification of averaging times for determining emission limit compliance.

The applicant should review the proposed conditions to determine whether day-to-day operation of the facility can be carried out in compliance. Particularly important are averaging times for determining emission compliance. Conditions can be expected for averaging times corresponding to the averaging times associated with ambient air quality standards (see Chapter 1), but imposition of conditions with other averaging times could result in operating problems, particularly during startup or upsets.

The wording of permit conditions is subject to negotiation with the permitting agency. The applicant should make the effort to negotiate these conditions as early in the application review process as they are being developed so that wording that will make compliance difficult or impossible can be avoided.

X. CONCLUSIONS

Developing and following a permitting strategy can make a significant difference in the time it takes to obtain a permit, the cost of obtaining the permit, and the success that can result once the permit is issued. Both careful planning and execution of the plan flexibly to accommodate unexpected changes in permitting agency requirements, agency staff, and project elements are needed to assure permitting success.

REFERENCES

1. Clean Air Act Amendments of 1990, Title V, Section 502(b)(6) and Section 503(e).
2. PL 90-542, 82 Stat 906, 16 USC 1271 ff.
3. United States Forest Service, 1977 "National Forest Landscape Management," Vol. 2, Chapter 1, The Visual Management System, in Agriculture Handbook 462, USDA.
4. BLM, 1980, *Visual Resource Management Program*.
5. PL 93-205, 87 Stat 884, 16 USC 1531 ff. with regulations at 50 CFR Section 17.1 *et seq.*
6. PL 89-665, 80 Stat 915, 16 USC 470 *et seq.* with regulations at 36 CFR Part 800.
7. Executive Order 11593 (May 13, 1971), 16 USC 470 with regulations at 36 CFR 8921.
8. 43 USC, Annotated Section 1996.
9. PL 101-601.
10. PL 91-190, 42 USC 4321 ff.
11. Section 102.
12. The Clean Air Act of 1977 excluded Prevention of Significant Deterioration permitting from NEPA review.
13. PL 99-499, 42 USC Sections 11001 et seq. with regulations at 29 CFR 1910 ff.
14. PL 94-5809, 90 Stat 95.
15. 42 USC 6901 *et seq.* with regulations at 40 CFR 260 *et seq.*
16. PL 96-510, 94 Stat 2767, 42 USC 9601 with regulations at 40 CFR Parts 300 ff.
17. PL 92-583, 16 USC 1451 ff.
18. PL 91-604, 42 USC 1857 ff.
19. PL 92-842, 42 USC 4901 ff.
20. Section 112(r) with regulations at 40 CFR Part 83.
21. PL 101-508.
22. Regulations are found at 49 CFR Subtitle B, Chapter I, subchapter C and Chapter III, subchapter B.
23. PL 92-500, 33 USC 1251 ff., 86 Stat 816.
24. PL 95-217, 91 Stat 1567.
25. PL 100-4, 101 Stat 76.
26. PL 93-523, 42 USC 300f, 88 Stat 1661 with regulations at 40 CFR Parts 122, 123, 146, and 147.
27. 33 USC 401-413.
28. PL 91-596, 84 Stat 1590 with regulations at 29 CFR 1900 *et seq.*
29. STAPPA/ALAPCO *Air Quality Permits — A Handbook for Regulators and Industry,* State and Territorial Air Pollution Program Administrators and the Association of Local Air Pollution Control Officials, 444 North Capitol Street, NW, Washington, DC 20001, 1991.
30. Claris Software.
31. For example, Microsoft Project® and Timeline®.

8

COMPLIANCE

I. COMPLIANCE REQUIREMENTS

Once a source has received an air quality permit, it is important to stay in compliance with all of the permit conditions included. The federal Title V program requires that each Title V permit specifies the regulations applicable to the permitted source. Furthermore, the Title V permit also specifies how compliance will be determined for each applicable regulation. The methods to determine compliance and the EPA Compliance Assurance Monitoring (CAM) program are discussed below.

A. Methods to Determine Compliance

As discussed in Chapter 1, Title V permit applications must identify both requirements applicable to the facility and any emission units and how compliance with those applicable requirements will be determined. The elements in demonstrating compliance include compliance monitoring, record keeping, and reporting.

1. Monitoring

Monitoring can include everything from continuous emission monitoring (CEM) to keeping records of complaints received. Monitoring can be quite expensive. Even if the equipment used for monitoring is not expensive, there is always a frequency of monitoring that must be met, and doing the monitoring takes staff resources that, for a large facility, can be substantial.

One of the most promising emission monitoring techniques recently developed is called Parametric Emission Monitoring System (PEMS).[1] PEMS uses a multivarient computer code to correlate data from parametric monitors on an emission source with measurements from a portable leased CEM system. During initial installation, information from all available parametric monitors is provided to the computer. The PEMS code then evaluates which of the parameters are independent and identifies those parameters. Long-term monitoring is then limited to the independent parameters. The PEMS code then correlates parameter changes with emission changes. Following testing in which the PEMS predicted emissions are compared to CEM data, the CEM system is removed, and PEMS provides emission estimates. Since parametric monitors such as fuel flow and temperature are usually much more reliable than CEM monitors, the availability of the PEMS is usually much higher than that of a CEM system. Operating costs also are lower, because daily calibrations and other CEM maintenance requirements are not necessary.

2. Opacity Monitoring

Opacity compliance must be addressed in virtually every Title V permit. Opacity limits are found both in agency prohibitory rules, in New Source Performance Standards, and in conditions for individual pieces of equipment. Several levels of opacity measurement can be considered, including:

- Continuous Opacity Monitoring Systems (COMS)
- EPA Method 9 Opacity Monitoring
- EPA Method 22 Opacity Monitoring
- Complaint Reporting and Response

a. Continuous Opacity Monitoring Systems. COMS are optical monitors at the top of a stack or vent that measure the attenuation of a light beam through stack gases. They provide a continuous reading of opacity that is recorded electronically. Although reliable, they require periodic cleaning of the optical elements to ensure that readings are of gas opacity and not extraneous contamination. Also, they are generally not able to distinguish between particulate matter and water vapor. Opacity due to water vapor is not considered a violation, so this equipment limitation can be important. One technique for distinguishing between particulate matter opacity and water vapor is to install the opacity monitor at a point in the stack where the temperature is above the condensation temperature of any moisture.

b. EPA Method 9 Monitoring. EPA Method 9 monitoring[2] is visual monitoring of emissions from a point source by a qualified observer or "smoke reader."* The certification is a course and examination offered by agencies or commercial schools, and to remain certified, a person must be recertified every 6 months. Monitoring consists of 24 observations of 15-sec duration and comparison of the opacity of the emissions to a standard gray scale. It is considered a violation if average opacity of the 24 measurements exceeds the standard.

For Method 9 to be used as a compliance demonstration, a facility must have a certified "smoke reader" available, and readings must be taken at intervals specified in permit conditions. Typically two employees are certified so that readings can be taken even when one of the certified persons is unavailable. One

---

* An alternate to Method 9 is also available using laser radar (LIDAR).

cement kiln that utilized Method 9 for demonstrating opacity compliance had 18 certified smoke readers on staff to ensure that readings were possible at any time during the 24-hr operation of the kiln. Notwithstanding, certified readings can only be taken during daylight hours when there is no precipitation.

Only COMS and Method 9 observations can be used to quantify opacity for comparison with a specified standard. The additional methods described below are qualitative.

c. EPA Method 22 Monitoring. EPA Method 22[3] is also a visual observation method, but it does not require a certified smoke reader. Method 22 suggests that observers be familiar with the techniques in Method 9, but asks only that an observer note the percentage of time visible opacity from emissions is observed during the observation period. The observation period must be at least 6 min in duration. The visible opacity must be sustained for the period specified in the applicable regulation to constitute a notable incident, and the requirement is that these incidents be documented, responded to, and reported.

d. Complaint Reporting and Response. The simplest method of opacity monitoring is to make a commitment to keep records of any opacity complaints received by a facility and to respond to those complaints with corrective action to eliminate observed opacity. Documentation of the corrective action taken must also be maintained. The record of complaints and corrective action is then provided to the agency as part of periodic reporting.

One method commonly used to check opacity is a combination of Method 22 and the complaint recording and response. The plant manager or other staff regularly drive around the plant to see if there are any visible emissions. For a continuously operated plant, one plant manager conducts this inspection upon arrival at the plant each day.

3. Record Keeping

Record keeping must be used to provide documentation of monitoring that has been conducted. For material balance compliance demonstrations, record keeping of material used and the composition of the material is the essence of the demonstration. For methods that use a surrogate parameter, or activity data, along with an emission factor to demonstrate compliance, records must be kept of the activity, such as fuel use or operating hours.

A standard condition for both construction and Title V permits is that permitting agency staff or their representatives have authority to enter the premises of the facility at "any reasonable time," and without prior notification, to inspect both monitoring methods and the records that demonstrate compliance.

Records need not be hard copy. Magnetic or other electronic recording is acceptable provided it can be displayed and audited. In some instances, agencies have required that data be available electronically from agency offices, so that compliance can be verified at any time. Although this requirement is unusual, it is well within the state of the art of current monitoring and data transfer systems.

The discipline of good record keeping is essential to demonstrating compliance with a Title V permit. Use of ditto marks or photocopying of record-keeping pages in a log indicate poor record-keeping discipline, and often are indicators of record-keeping errors. One advantage of electronic record keeping is that repeated records can be copied, and then appropriate changes made to update the inputs.

Periodic checking of records by a person unfamiliar with the operation can quickly identify entries that are understandable only to the equipment operator or person making record-keeping entries. The records, including the signature or initials of the person making the entries, must be able to be understood by others.

4. Reporting

For Title V permit holders, periodic reports must be provided on the results of compliance monitoring. Reporting for Title V permits is required either every 6 months or annually. Reports are public and can result in enforcement actions if noncompliance is reported.

5. Good Conduct Provisions

For any compliance monitoring that is included in a Title V permit, a good conduct provision should be considered. Such provisions have been included in NSPS. They specify that initially periodic monitoring be done frequently; if no noncompliance is monitored, the frequency of monitoring decreases. Several steps of decreases can be included, and if during any monitoring subsequent to the decrease in frequency noncompliance is observed, the monitoring steps back to a more frequent interval and begins the process again. This good-conduct provision provides an incentive for the facility to stay in compliance, thereby limiting the expense of periodic monitoring.

II. Clean Air Act Enforcement Provisions

There are several enforcement options that EPA can take under the CAA. These include:

- Issuing an administrative penalty order
- Issuing an order requiring compliance or prohibition
- Bringing civil action in court
- Requesting the Attorney General to bring criminal action

Each of these is discussed below. Individual states may have similar provisions for violation of state statutes and will have provisions at least as stringent as these for violation of state statutes that are delegation of federal statutes.

CIVIL PENALTIES CAN BE ASSESSED UNDER THE CLEAN AIR ACT

- Field citations: No more than $5,000 per day up to a maximum of $25,000
- Civil penalties: Up to $25,000 per day for each violation
- Whistleblower awards of $10,000 can be given by EPA
- Recovery of attorney and expert witness fees can be assessed

---

A. Administrative Penalty Orders

Administrative penalty assessments may be made for violations of any of the provisions of the CAA, including violation of permit conditions. Field citations can be issued by an inspector. These citations have a fine of $5,000 per day, and they may accumulate up to a maximum of $25,000. Individuals receiving field citations may request a hearing if they do so within a reasonable time; they will have opportunity to present evidence in defense.

Civil penalties can be assessed for up to $25,000 per day for each violation. Also under the category of administrative penalties are whistleblower awards of up to $10,000 that can be made by EPA to persons reporting a violation. A person receiving an administrative penalty may appeal the assessment within 30 days, or may appeal a final ruling to the federal district court in an additional 30-day window. The Administrator of EPA is authorized to compromise, modify, or place conditions on a penalty.

Persons receiving administrative penalties should promptly contact the regulating agency to clarify any uncertainty about the violation, identify extenuating circumstances, indicate that the cited noncompliance has been corrected, and negotiate reductions in the penalty. Extenuating circumstances are considered by the agency, and it is unusual for a final penalty to remain as high as the original amount.[4]

B. Issuing an Order Requiring Compliance or Prohibition

Issuing an order is a more formal, and more serious, response to a violation. It is usually used when a specific problem must be addressed, and when there is the potential for impairing human health or of serious or irreversible damage to the environment. A hearing is required to address an order, and civil or criminal penalties may be assessed subsequently.

C. Bringing Civil Action in Court

Civil action enforcement is more serious, and potentially a great deal more expensive than the administrative actions discussed above. There is provision in the statute for recovery of attorney and expert witness fees from the defendant during civil action. Under Title VII of the Clean Air Act Amendments of 1990, citizen suits can be brought as civil action. Such suits can be based on a Title V permit application that identifies current noncompliance, even if a compliance plan is included in the application. The exception to the ability to bring a citizen suit against current noncompliance is if the noncompliance has already been adjudicated and the compliance plan is part of a consent decree already approved by the court.

D. Requesting the Attorney General to Bring Criminal Action

Provisions under federal law as well as laws in many states provide for criminal sanctions against both organizations and individuals. Usually, criminal action is reserved for flagrant, intentional violations. It includes up to 2 years imprisonment for false statements or noncompliant monitoring or record keeping, 1 year for negligent endangerment, and up to 15 years and a fine of up to $1,000,000 for knowing endangerment of death or serious bodily injury.[5]

THE CLEAN AIR ACT ALSO PROVIDES FOR CRIMINAL ENFORCEMENT INCLUDING IMPRISONMENT

- Two years for false statements, non-compliant monitoring and record keeping
- 1 year for negligent endangerment
- 15 years for knowing endangerment including knowing release of HAP or EHS causing imminent danger
- 5 years for knowing violation of SIP, NSPS, NESHAP, Title V, or Section 114 requirements
- Maximum fine of $250,000 for individuals, $500,000 for organizations

---

E. Emergency as a Defense

If there is a release of pollutants that would otherwise be considered a violation, it is possible to argue that it was an emergency. For this purpose, an emergency is defined as any sudden and reasonably unforeseeable event beyond the control of the source that requires immediate corrective action to restore normal operation, and causes exceedence of technology-based emission limitations.

If the noncompliance is caused by improperly designed equipment, lack of preventative maintenance, careless or improper operation, or operator error, it is not considered an emergency. An emergency can be used as a defense if there are properly signed operating logs or other evidence that indicates an emergency occurred and the cause can be identified, the facility was being properly operated at the time, and the permittee took all reasonable steps to minimize levels of exceedences.

To use this defense, notice must be provided to the permitting agency within two working days containing a description of the emergency and steps taken to mitigate emissions and perform corrective actions. If the only way to reduce the emissions to allowable levels is to halt or reduce activity at the plant, and that is not done, the emergency defense cannot be used. That is, if you do not shut down, you are liable.

As with other compliance issues, good record keeping is vital to protection from prosecution when using the emergency defense. It is necessary to have records indicating that the plant had been operating normally and that efforts were made to halt or reduce the excess emissions.

F. Section 114 Fact-finding

EPA has the authority[6] to obtain information from a source in anticipation of prosecution. If a source receives a letter from EPA citing Section 114, it should be considered at least as serious as an Internal Revenue Service audit. EPA may ask questions or seek information about both emissions and the structure of environmental management in the company. EPA may request records, monitoring, sampling, or compliance certifications. EPA has the right of entry to company premises on which records are kept when credentials are presented.

SECTION 114 IS FACT-FINDING IN ANTICIPATION OF PROSECUTION

- EPA may request records, monitoring, sampling, or compliance certifications.
- Receipt of a Section 114 letter is equivalent to an IRS audit.
- Questions about both emissions and the way environmental management is organized are likely.
- Section 114 responses should be prompt, accurate, and limited to the questions asked.

Often a 114 letter is stimulated by evidence obtained by EPA that excess emissions have occurred. This can include emissions reported as part of EPCRA 313 Form R reporting of releases to the environment, newspaper accounts, or notices of receipt of a variance from state emission limits.

Responses to Section 114 letters should be prompt, accurate, and limited to the questions asked. Counsel should participate in preparing or reviewing the response and in any subsequent meetings with the agency staff.

G. Inspection Protocol

Inspections are a standard activity for permitting agency staff, and a facility should expect to be inspected periodically. Some very large facilities, such as oil refineries, have a full-time inspector from the regulating agency assigned, and this person is at the facility every day. Other facilities have less frequent inspections. For any facility, an inspection protocol is worthwhile.

The inspection protocol should identify who is to respond when an inspector comes, what responses are appropriate, and what rights the inspector has as well as rights the company has.

The agency staff, or an authorized representative, has the right of entry to a facility. The receptionist or gate guard should be instructed about whom to call when an inspector presents his or her credentials and asks for entry. A member of the environmental staff should greet the inspector and, unless the inspection is anticipated and routine, take the inspector to a conference room for a preinspection conference. It is appropriate to ask the inspector what he or she wants to see. Although the company cannot bar access, safety provisions are often necessary for visiting certain areas of the plant, and planning such visits is usually required by plant safety rules. Inspectors should be required to use all safety equipment required of employees and other visitors, and in some instances, a safety briefing is needed before visiting the plant.

Once the area to be visited or information sought has been identified, the inspector should be escorted promptly and directly to that area. It can be a safety hazard as well as present a risk of providing misleading information for an inspector to visit a facility without an escort. In the event that the inspector requests a stop at another area not covered by the original request, such a stop can be prohibited until another initial interview has been conducted, since the request was not covered in the original interview.

If an inspector uses a recorder or any test instruments, the company should also record the same information and test the same location. In some instances, it may be appropriate to videotape the inspection.

At the end of the inspection, an exit conference should be held. The inspector should be asked what information was obtained, and copies of any information should be requested.

Most companies develop good working relationships with air quality inspectors who regularly visit their plants, and it is important that procedures used to respond to an inspection are not hostile or unnecessarily inhibiting. However, it is not unusual for an inspector, after working some time with a facility, to assume that he or she is as knowledgeable of facility operation as the plant operators. Since this is almost never the case, escorts and explanations of operations are needed, even for experienced inspectors.

REFERENCES

1. Keeler, J. et al., "Achieving Compliance and Profits with a Predictive Emissions Monitoring System: Pavillions Software CEM™, "Pavillions Technologies, Inc., 3500 W. Balcones Center Dr., Austin, TX 78759.
2. 40 CFR Part 60, Appendix A Method 9 Visual Determination of the Opacity of Emissions From Stationary Sources.
3. An alternate to Method 9 is also available using laser radar (LIDAR).
4. 40 CFR Part 60, Appendix A Method 22 Visual Determination of Fugitive Emissions From Material Sources and Smoke Emissions From Flares.
5. Anglehart, Don, "Take 'Penalty Policies' with a Grain of Salt," Business & Legal Reports, Inc., EM378, p.7, 1994.
6. Clean Air Act Section 113 (c).
7. Clean Air Act Section 114.

1.a. The annual consumption of natural gas will be the product of the hours of operation on gas times the heat rate of the boiler divided by the heating value of natural gas:

Annual gas consumption =

$$\frac{\left[\left(50\frac{week}{year}\right)\left(7\frac{days}{week}\right)\left(24\frac{hour}{day}\right)-10\frac{hour}{month}\times12\frac{month}{year}\right]\left(60\times10^6\frac{Btu}{hour}\right)}{1000\frac{Btu}{ft^3}}$$

$$=4.968\times10^8\frac{ft^3}{year}=497\frac{million\ ft^3}{year}$$

Annual diesel consumption =

$$\frac{\left(10\frac{hour}{month}\times12\frac{month}{year}\right)\left(60\times10^6\frac{Btu}{hour}\right)}{\left(130,000\frac{Btu}{gallon}\right)}$$

$$=55,384\frac{gallons}{year}=55.3\frac{thousand\ gallons}{year}$$

b. The NO_x emission rate will be the emission factor for NO_x times the annual fuel use totaled for both fuels.

NO_x Emission Rate

$$= \left[\left(100\frac{\text{lb}}{\text{million ft}^3}\right)\left(497\frac{\text{million ft}^3}{\text{year}}\right)\right]_{\text{gas}} + \left[\left(20\frac{\text{lb}}{10^3\text{gallon}}\right)\left(55.3\frac{10^3\text{gallon}}{\text{year}}\right)\right]_{\text{oil}}$$

$$= \left(49,700\frac{\text{lb}}{\text{year}}\right)_{\text{gas}} + \left(1106\frac{\text{lb}}{\text{year}}\right)_{\text{oil}} = 50,806\frac{\text{lb}}{\text{year}} = 25.4\frac{\text{ton}}{\text{year}}$$

c. The cadmium emission rate will be the emission factor for cadmium times the annual oil use, since there is no cadmium in natural gas.

$$\text{Cd Emission Rate} = \left(0.0015\frac{\text{lb}}{10^3\text{gallon}}\right)\left(55.3\frac{10^3\text{gallon}}{\text{year}}\right) = 0.083\frac{\text{lb}}{\text{year}}$$

d. The potential to emit for cadmium from this boiler will be the emission factor for cadmium times the maximum oil use possible for the boiler — if the boiler operated on oil throughout the year.

$$\text{Cadmium PTE} = \left(0.0015\frac{\text{lb}}{10^3\text{gallon}}\right)\frac{\left(8760\frac{\text{hour}}{\text{year}}\right)\left(60\times10^6\frac{\text{Btu}}{\text{hour}}\right)}{130,000\frac{\text{Btu}}{\text{gallon}}} = 6.1\frac{\text{lb}}{\text{year}}$$

2. The change in emissions will be the emission factor times the change in operating hours for the boiler.

$$\Delta\text{ Emissions (NO}_x) = \frac{\left[(5,000-120)\frac{\text{hour}}{\text{year}}\right]\left(56\frac{\text{million Btu}}{\text{hour}}\right)}{130,000\frac{\text{Btu}}{\text{gallon}}}\left(20\frac{\text{lb}}{10^3\text{gallon}}\right)$$

$$= 42,043\frac{\text{lb}}{\text{year}} = 21\frac{\text{ton}}{\text{year}}$$

$$\Delta\text{ Emissions (SO}_2) =$$

$$\frac{\left[(5000-120)\frac{\text{hour}}{\text{year}}\right]\left(56\frac{\text{million Btu}}{\text{hour}}\right)}{130,000\frac{\text{Btu}}{\text{gallon}}}\left(142\frac{\text{lb}}{10^3\text{gallon }\%\text{ sulfur}}\right)(0.05\%\text{ sulfur})$$

$$= 14,925\frac{\text{lb}}{\text{year}} = 7.5\frac{\text{ton}}{\text{year}}$$

$$\Delta \text{Emissions (CO)} = \frac{\left[(5000-120) \dfrac{\text{hour}}{\text{year}} \right] \left(56 \dfrac{\text{million Btu}}{\text{hour}} \right)}{130,000 \dfrac{\text{Btu}}{\text{gallon}}} \left(5 \dfrac{\text{lb}}{10^3 \text{gallon}} \right)$$

$$= 10,511 \frac{\text{lb}}{\text{year}} = 5.3 \frac{\text{ton}}{\text{year}}$$

b. From Table 1-3, this change does not require a major modification because it is less than any of the applicable major modification thresholds for a serious nonattainment area.

3.a. The emission factors for these boilers will depend on their design capacity. Since the design capacity in Table 2-1 is in units of million Btu/hr, the first step will be to determine the design capacity in those units.

$$\text{Design capacity} = \left(150,000 \frac{\text{ft}^3}{\text{hr}} \right) \left(1000 \frac{\text{Btu}}{\text{ft}^3} \right) = 150 \frac{10^6 \text{Btu}}{\text{hour}}$$

The boilers are in the utility boiler size range. Although the emission factor in Table 2-1 for a utility boiler is $550 \dfrac{\text{lb}}{10^6 \text{ft}^3}$ of natural gas, this large an emission rate would not be allowed under the New Source Performance Standard for utility boilers. From Figure 2-2, the allowable emission factor for a utility boiler with a high heat rate (>70,000 $\dfrac{\text{Btu}}{\text{hr ft}^3}$) is $200 \dfrac{\text{lb}}{10^6 \text{ft}^3}$. The potential to emit for this plant will be the design capacity times the maximum hours of operation for the plant per year times the emission factor.

Potential to Emit (NO_x)

$$= \frac{2 \left(150 \dfrac{10^6 \text{Btu}}{\text{hour}} \right) \left(8760 \dfrac{\text{hour}}{\text{year}} \right)}{1000 \dfrac{\text{Btu}}{\text{ft}^3}} \left(200 \dfrac{\text{lb}}{10^6 \text{ft}^3} \right) = 525,600 \frac{\text{lb}}{\text{year}} = 263 \frac{\text{ton}}{\text{year}}$$

Potential to Emit (SO_2)

$$= \frac{2 \left(150 \dfrac{10^6 \text{Btu}}{\text{hour}} \right) \left(8760 \dfrac{\text{hour}}{\text{year}} \right)}{1000 \dfrac{\text{Btu}}{\text{ft}^3}} \left(0.6 \dfrac{\text{lb}}{10^6 \text{ft}^3} \right) = 1577 \frac{\text{lb}}{\text{year}} = 0.8 \frac{\text{ton}}{\text{year}}$$

Potential to Emit (PM_{10})

$$= \frac{2\left(150\,\dfrac{10^6\,\text{Btu}}{\text{hour}}\right)\left(8760\,\dfrac{\text{hour}}{\text{year}}\right)}{1000\,\dfrac{\text{Btu}}{\text{ft}^3}}\left(5\,\dfrac{\text{lb}}{10^6\,\text{ft}^3}\right) = 13{,}140\,\dfrac{\text{lb}}{\text{year}} = 6.6\,\dfrac{\text{ton}}{\text{year}}$$

Potential to Emit (CO)

$$= \frac{2\left(150\,\dfrac{10^6\,\text{Btu}}{\text{hour}}\right)\left(8760\,\dfrac{\text{hour}}{\text{year}}\right)}{1000\,\dfrac{\text{Btu}}{\text{ft}^3}}\left(40\,\dfrac{\text{lb}}{10^6\,\text{ft}^3}\right) = 105{,}120\,\dfrac{\text{lb}}{\text{year}} = 53\,\dfrac{\text{ton}}{\text{year}}$$

b. To be constructed this plant will need to obtain a PSD permit for pollutants that are in attainment within the impact area of the plant and an NSR permit for nonattainment pollutants. The plant will also need to obtain a Title V operating permit.
4. The emissions will be the emission factor for an institutional boiler for each pollutant times the hours of operation of the boiler. Since no hours of operation are specified, permitting will need to be based on continuous operation, 8760 hours per year.

Potential to Emit (NO_x)

$$= \frac{\left(6.5\,\dfrac{10^6\,\text{Btu}}{\text{hour}}\right)\left(8760\,\dfrac{\text{hour}}{\text{year}}\right)}{1000\,\dfrac{\text{Btu}}{\text{ft}^3}}\left(100\,\dfrac{\text{lb}}{10^6\,\text{ft}^3}\right) = 5694\,\dfrac{\text{lb}}{\text{year}} = 2.8\,\dfrac{\text{ton}}{\text{year}}$$

Potential to Emit (VOC)

$$= \frac{\left(6.5\,\dfrac{10^6\,\text{Btu}}{\text{hour}}\right)\left(8760\,\dfrac{\text{hour}}{\text{year}}\right)}{1000\,\dfrac{\text{Btu}}{\text{ft}^3}}\left(8\,\dfrac{\text{lb TOC}}{10^6\,\text{ft}^3}\right)\left(0.48\,\dfrac{\text{lb VOC}}{\text{lb TOC}}\right) = 219\,\dfrac{\text{lb}}{\text{year}} = 0.1\,\dfrac{\text{ton}}{\text{year}}$$

Potential to Emit (SO_2)

$$= \frac{\left(6.5\,\dfrac{10^6\,\text{Btu}}{\text{hour}}\right)\left(8760\,\dfrac{\text{hour}}{\text{year}}\right)}{1000\,\dfrac{\text{Btu}}{\text{ft}^3}}\left(0.6\,\dfrac{\text{lb}}{10^6\,\text{ft}^3}\right) = 34.16\,\dfrac{\text{lb}}{\text{year}} = 0.02\,\dfrac{\text{ton}}{\text{year}}$$

Potential to Emit (PM_{10})

$$= \frac{\left(6.5\dfrac{10^6\,\text{Btu}}{\text{hour}}\right)\left(8760\dfrac{\text{hour}}{\text{year}}\right)}{1000\dfrac{\text{Btu}}{\text{ft}^3}}\left(12\dfrac{\text{lb}}{10^6\,\text{ft}^3}\right) = 683\dfrac{\text{lb}}{\text{year}} = 0.34\dfrac{\text{ton}}{\text{year}}$$

Potential to Emit (CO)

$$= \frac{\left(6.5\dfrac{10^6\,\text{Btu}}{\text{hour}}\right)\left(8760\dfrac{\text{hour}}{\text{year}}\right)}{1000\dfrac{\text{Btu}}{\text{ft}^3}}\left(21\dfrac{\text{lb}}{10^6\,\text{ft}^3}\right) = 1,196\dfrac{\text{lb}}{\text{year}} = 0.6\dfrac{\text{ton}}{\text{year}}$$

5.a. Using a mass balance approach, the emissions of MIBK will be the quantity of MIBK in the paint since there are no devices to control MIBK emissions.

Emissions (MIBK)

$$= \left(4000\dfrac{\text{gal}}{\text{year}}\right)\left(3.85\dfrac{\text{lb MIBK}}{\text{gal}}\right) = 15,400\dfrac{\text{lb MIBK}}{\text{year}} = 7.7\dfrac{\text{ton}}{\text{year}}\text{ MIBK}$$

b. The total VOC emissions from this source will also be 7.7 ton/year, since acetone has been removed by EPA from the list of compounds that are considered VOCs.
c. Particulate emissions will be the quantity of solids in the paint minus the solids transferred to the part minus the solids captured in the mat filters.

Emissions (PM_{10})

$$= \left(4000\dfrac{\text{gal}}{\text{year}}\right)(7-3.85-1.4)\dfrac{\text{lb}}{\text{gal}}(1-0.65)(1-0.9) = 245\dfrac{\text{lb}}{\text{year}} = 0.12\dfrac{\text{ton}}{\text{year}}$$

6.a. VOC emissions will be the quantity of VOC in the paint minus the VOC removed in the incinerator.

Emissions (VOC)

$$= \left(1400\dfrac{\text{gal}}{\text{year}}\right)(450\text{ g/L})\left(\dfrac{3.85\text{ L/gal}}{454\text{ g/lb}}\right)(1-0.95) = 267\dfrac{\text{lb}}{\text{year}} = 0.13\dfrac{\text{ton}}{\text{year}}$$

b. Ozone precursors include the VOC emissions from the paint and the NO_x emissions from the incinerator.

Emissions (Precursors)

$$= \left(0.113 \frac{\text{ton}}{\text{year}} \right)_{VOC} + \left[\left(0.6 \frac{\text{mmscf}}{\text{day}} \right) \left(240 \frac{\text{day}}{\text{year}} \right) \left(300 \frac{\text{lb}}{\text{mmscf}} \right) \left(\frac{\text{ton}}{2000 \text{ lb}} \right) \right]_{NO_x}$$

$$= 21.7 \frac{\text{ton}}{\text{year}}$$

c. Hexavalent chromium emissions will be the quantity of hexavalent chromium in the paint minus the amount applied minus the amount captured in the mat filters.

Emissions (CrVI)

$$= \left(1400 \frac{\text{gal}}{\text{year}} \right) \left(7 \frac{\text{lb}}{\text{gal}} \right) (0.0001)(1 - 0.65)(1 - 0.80) = 0.07 \frac{\text{lb}}{\text{year}}$$

7. The emissions of TCE are the difference between the quantity of TCE coming into the stripper and the quantity leaving.

Emissions (TCE)

$$= \left(2500 \frac{\text{gal}}{\text{day}} \right) \left(3.85 \frac{\text{L}}{\text{gal}} \right) (0.370 - 0.015) \frac{\text{mg}}{\text{L}} = 3417 \frac{\text{mg}}{\text{day}} \left(365 \frac{\text{day}}{\text{year}} \right) \left(\frac{\text{lb}}{454,000 \text{ mg}} \right)$$

$$= 2.7 \frac{\text{lb}}{\text{year}}$$

8. The benzene emission rate will be the sum of the emissions from each component type. The emissions from each component type will be the number of components times the emission factor for that component.

$$\text{Emission (Pumps)} = (12 \text{ pumps}) \left(0.0494 \frac{\text{kg}}{\text{hr} - \text{source}} \right) = 0.593 \frac{\text{kg}}{\text{hr}}$$

$$\text{Emission (Valves)} = (140 \text{ valves}) \left(0.0071 \frac{\text{kg}}{\text{hr} - \text{source}} \right) = 0.994 \frac{\text{kg}}{\text{hr}}$$

$$\text{Emission (Flanges)} = (435 \text{ flanges}) \left(0.00083 \frac{\text{kg}}{\text{hr} - \text{source}} \right) = 0.361 \frac{\text{kg}}{\text{hr}}$$

$$\text{Emission (Sample)} = (15 \text{ samples})\left(0.0150 \frac{\text{kg}}{\text{hr} - \text{source}}\right) - 0.225 \frac{\text{kg}}{\text{hr}}$$

Total Emissions

$$= (0.593)_{\text{pump}} + (0.994)_{\text{valve}} + (0.361)_{\text{flange}} + (0.225)_{\text{sample}}$$

$$= \left(2.173 \frac{\text{kg VOC}}{\text{hour}}\right)\left(8760 \frac{\text{hour}}{\text{year}}\right)\left(\frac{2.2 \text{ lb}}{\text{kg}}\right) = 41,878 \frac{\text{lb VOC}}{\text{year}}$$

Emissions (Benzene)

$$= \left(41,878 \frac{\text{lb VOC}}{\text{year}}\right)\frac{[(5.5)(0.21) + (6.5)(.017)]}{12} = 789 \frac{\text{lb}}{\text{year}}$$

9.a. The concentration of lead in this stack will be the quantity of lead found in the sample divided by the volume of the sample.

$$\text{Concentration (Pb)} = \frac{60 \text{ mg}}{4 \text{ m}^3} = 15 \frac{\text{mg}}{\text{m}^3}$$

b. The emission rate of lead will be the concentration times the flow rate.

Emission (Pb)

$$= \left(15 \frac{\text{mg}}{\text{m}^3}\right)\left(800 \frac{\text{m}^3}{\text{min}}\right)\left(60 \frac{\text{min}}{\text{hour}}\right)\left(8760 \frac{\text{hour}}{\text{year}}\right)\left(\frac{\text{lb}}{454,000 \text{ mg}}\right)$$

$$= 13,893 \frac{\text{lb}}{\text{year}} = 6.9 \frac{\text{ton}}{\text{year}}$$

INDEX

A

AB 2588, 116
abundantly cautious, 141
Acceptable Exposure Level (AEL), 129
activity data, 36
acute health effects, 115, 125, 129–130,
      133–134
administrative penalty order, 164–165
aesthetics, 145, 149
agency, 28, 138–140, 142–143, 145, 151, 153,
      155, 160, *see also* permitting
      strategy
air basin, 1
Air Pollution Control Officer (APCO), 67, 70
Air Quality Impact Analysis (AQIA), 17–19,
      93–113
Air Quality Related Values (AQRV), 3, 17, 19
alternate operating scenarios, 26
ambient air quality, 3–9, 14, 17, 25
ambient air quality standards, 3, 97, 109–110,
      112
ambient monitoring, 107
American Indian Religious Freedom Act of
      1978, 146, 149
Antiquities Act of 1906, 147, 151
AP-42, 38–42, 116
area sources, 9, 35, 101
attainment area, 7, 8, 11, 12, 14, 19, 21
averaging time, 4, 159

B

base realignment and closure (BRAC), 86
benzene, 115, *see also* emission factor
best available control measure (BACM), 66, 69
Best Available Control Technology (BACT), 2,
      17, 20, 65–80, 142, 153–155
  achieved in practice, 66, 67, 71

annual cost, 73–75, 79
available, 70–73
boiler, 67, 77–79
capital costs, 73–75, 78–79
compared to MACT, 122
cost effectiveness, 73–76, 78
economic feasibility, 67–68
economic impacts, 65, 70, 73
energy, 65, 68, 70, 73, 78
environmental, 65, 68, 70, 73, 78
feasibility, 67, 70–72, 77
general and administrative costs (G&A),
      76
infeasible, 70, 72, 73
NO_x, 68–73, 76, 77–79
operating and maintenance cost (O&M), 76,
      79
protocol, 70
strategy, 139–140, 142, 153–155
top-down, 66, 69–73
threshold, 2
best available retrofit control technology
      (BARCT), 66, 68
biological resources, 145, 149
Bureau of Land Management (BLM), 145, 151

C

California Air Resources Board (CARB), 48, 68
cancer, 115, 123, 126, 128–131
  burden, 130, 132–133
  potency factors, 129–130
  slope factor (SF), 129–130
capital cost, 139, *see also* Best Available
      Control Technology (BACT)
capital recovery factor (CRF), 75
carbon monoxide (CO), 4, 76
chronic health effect, 115, 129, 130, 133
citizen suit, 27, 41

O

P